Roll-to-Roll Vacuum Deposition of Barrier Coatings

Scrivener Publishing
3 Winter Street, Suite 3
Salem, MA 01970

Roll-to-Roll Vacuum Deposition of Barrier Coatings

Charles A. Bishop
C.A. Bishop Consulting Ltd.,
Leicestershire, UK

Co-published by John Wiley & Sons, Inc. Hoboken, New Jersey, and Scrivener Publishing LLC, Salem, Massachusetts.
Published simultaneously in Canada.

For general information on our other products and services or for technical support, please contact our Customer Care Department within the United States at (800) 762-2974, outside the United States at (317) 572-3993 or fax (317) 572-4002.

Wiley also publishes its books in a variety of electronic formats. Some content that appears in print may not be available in electronic formats. For more information about Wiley products, visit our web site at www.wiley.com.

For more information about Scrivener products please visit www.scrivenerpublishing.com.

Cover design by Russell Richardson.

Library of Congress Cataloging-in-Publication Data:

ISBN 978-0-470-60956-9

Printed in the United States of America

10 9 8 7 6 5 4 3 2 1

Contents

Preface

Barrier materials have been used from time immemorial such as to protect items from different environmental aspects such as sunlight, air, moisture and dirt. An early food packaging application was food wrapped in leaves or animal skins and buried in the earth. The leaves protected the food from the earth but also once overlapped and compressed by the earth helped keep moisture and oxygen away from the food too. The combination of earth and leaves also provided a light barrier. As technology progressed pottery, glass and tin cans all found use as barrier containers allowing food to be preserved for much longer times and more conveniently than by burying everything.

In more modern times we have had the development of polymers which has challenged many of the more traditional packaging materials. The polymers offer light weight and a versatility that means they have replaced many materials in a wide range of applications. However they could be even more widely used if their barrier performance could be improved.

Roll-to-roll vacuum deposition has been used to deposit the metal or glass-like coatings onto polymers for many years for barrier applications. These coatings have successfully been used for food packaging although there has always been pressure to improve the barrier and reduce the costs. More recently the requirements changed radically because of the electronics industry where to protect various products the barrier required needed to be a million times better than those used in food packaging.

Glass and metal foil have a proven very high barrier performance and so the simple belief was to use a combination of glass and polymer or metal and polymer to get the best of both worlds. Unfortunately there is a gap between theory and practice. The barrier performance is improved but nowhere near as much as had been predicted. Thus there has been a race to find ways to improve the basic vacuum deposition process, capable of producing the food packaging barrier materials, such that these ultra-barrier performance materials could be produced.

Vacuum deposition is a vast subject as too is barrier packaging. The aim of this book is to collect together the relevant information and references to make it easier for those wanting to vacuum deposit barrier coatings of all sorts to find out the information they need. It is not aimed to delve deeply into the physics of the permeation process and so I have tried to keep the mathematics to a minimum. I aim to give only the equations that are of practical help, either to evaluate barrier coatings or to develop barrier packages.

The use of vacuum deposited barrier coatings is determined by their performance and this can vary over more than six orders of magnitude. However the same principles apply to the manufacture of all vacuum deposited barrier coatings and the limitation of the barrier performance of the coatings is governed by the same factors. The barrier performance is primarily defined by the defects in the coating and minimising the defects is the primary goal at all levels of manufacturing. The highest performing barrier coatings are known as ultra barrier coatings and any defects are disastrous to these coatings. However even for food packaging coatings where the barrier performance is six orders of magnitude worse than for the ultra barrier coatings the same defects will also degrade the barrier performance.

The management of defects starts not with the deposition process but with the supply of materials, in particular the web substrates. The web substrate is a process variable and a critical one and therefore one that has to be managed.

This book is aimed at sharing much of the learning that occurred about why the coatings fail to achieve the theoretical performance and what might be done to produce improvements.

Whilst it is hoped that everything is explained clearly it is likely that there will be points that need clarification. If anyone wants to ask questions please feel free to contact the author either via the website at www.cabuk1.co.uk or via the blog website at www.vacuumcoatingblog.com. Similarly if you have new information, products, examples, experiences of different problems, etc that you think would improve the book for next time around, I would be happy to hear about them.

Charles A. Bishop
January 28, 2010

Acknowledgements

Any book on barrier coatings would not be complete without acknowledging having a debt of gratitude to the late Dr Bernard Henry. Bernard researched and championed the topic of barrier coatings for a number of years until his untimely death in 2007 at the early age of 42. So many of us all over the world were used to having Bernard around to be able to call or e-mail and discuss with him details of barrier coatings or applications and I am sure I am one amongst many who miss those interactions. Had he been still with us I am sure he would have taken up the challenge of writing the definitive book about vacuum depositing barrier coatings and I only hope that this attempt does the topic justice.

1

Introduction

To start let us explain what is meant when we refer to a barrier material and then go on to describe where these barrier materials might be used. A barrier is something that provides a resistance against something else so that in packaging or encapsulation terms the meaning of barrier tends to mean as resistance to the ingress of something that might degrade the product being packaged or encapsulated.

Everyone has their own idea of what is meant by a barrier material depending on what is most detrimental to their product. For some people barrier only refers to protection against water vapour whereas for others it is more important to protect the goods against oxygen. So barrier is a generic term that needs to be elaborated as a good barrier material against water vapour may have little barrier performance against oxygen and vice versa.

As will be shown there is a growing market for these existing barrier materials. There is also an, as yet, unexploited market for barrier materials with a much better barrier performance. As the performance requirements increase the difficulty in producing the barrier materials also increases, as does the cost.

The barrier performance starts with the choice of materials and includes the whole manufacturing process. Glass bottles and tin cans are long established barrier materials for food packaging. It was believed that to improve the barrier performance of polymers it was simply a matter of adding a thin enough glass or metal layer that would not impair the flexibility but, it was thought, would match the performance of the bottles or cans. The standard method of applying a glass or metal coating to polymer substrates is by vacuum deposition. In roll-to-roll vacuum deposition the quality of the supply rolls is a critical factor as too is any pre-treatment or cleaning process. Subsequent chapters will follow the process through from the polymer web production and any cleaning or pre-treatment through to the nucleation and growth of coatings deposited by various vacuum deposition techniques. In order to be able to compare the performance of different barrier coatings it is necessary to be able to measure the performance and so there is also a chapter that describes the most common methods of measuring the barrier performance. In this way it is hoped that it can be shown how the ultimate performance of the barrier materials can be affected throughout the whole manufacturing process.

A barrier is anything that keeps things apart and we can see examples of barrier materials everyday in food packaging where food products are protected from a variety of different elements be they gases, liquids or solids. Depending on the food and the sensitivity of the foods to degradation they may need protecting from moisture, oxygen, light as well as bacteria, moulds, aromas and taint (1). As might be expected different materials will perform differently as a barrier to liquids or gases and so there is not any one material used as a universal barrier. There are many possible solutions to providing a suitable barrier in fact one of the problems we now have is the vast choice of materials that in combination could provide the necessary barrier performance.

It is not just food that requires barrier materials but anything that has some sensitivity to the ingress or egress of some other material, be it a gas or moisture, will require a barrier to protect it. Thermal insulation panels, used to improve the insulation performance of buildings, are designed to have a working lifetime of 50–100 years. Throughout this time these panels are expected to maintain their insulation performance which is, in part, dependent upon the evacuated panel remaining under vacuum and hence air and moisture have to be kept out for all this time. The reality is that this is not achieved by the barrier material performance alone but by a combination of the barrier material and scavenger materials incorporated into the product that getter what little amount of gas or moisture that is passed through the barrier material. Once the scavenger material is saturated there will be a build up of gas or moisture and the performance of the insulation will then begin to decline.

In the area of electronics there are the organic light emitting devices (OLEDs) that are degraded by moisture ingress and are so sensitive to attack that the barrier requirements are six orders of magnitude higher than those used in most food packaging applications. These very high performance barrier materials are often referred to as ultra-barrier materials.

It is interesting to note that often the same materials are used for both the food packaging and for the ultra-barrier applications. There can be a huge performance difference for exactly the same materials that is dependent upon the quality of how the materials are supplied, handled and used to make the final barrier material. Polymer webs have a certain amount of barrier performance that is inherent but it is often not enough to meet the customer specifications and so is coated with something to improve the barrier performance. The two materials that have been used for food packaging for decades are metal and glass. The expectation being that adding a very thin glass or metal layer would change the polymer barrier performance into the same perfect performance exhibited by the glass or metal. The metal and glass or glass-like very thin layers, sometimes as thin as a few nanometers, can be deposited using vacuum deposition techniques. The question that has taken time to answer is what happens to these materials when they are deposited as very thin coatings that they no longer perform as well as when they are in the more rigid thick form.

Vacuum deposition onto flexible webs is where a roll of material is loaded onto a winding mechanism that is enclosed in a vacuum vessel that can be evacuated to remove the air. Different materials can be evaporated, or deposited by a variety of means onto the web as it is wound between unwind and rewind rolls. The lack of air enables metals to be deposited with minimal oxidation or for controlled stoichiometry compounds to be deposited. Glass as used in packaging is very rigid but if the glass is thinned down it shows increasing flexibility. The very thin glass used for displays that is less than 500 microns thick can be flexed and bent without breaking. If this same glass is vacuum deposited onto a flexible polymer web at a thickness of less than 15 nm the glass becomes even more flexible making it suitable for use in flexible packaging applications.

Similarly metals are also much more flexible when vacuum deposited as thin films than when produced as a rolled thin foil. Aluminium foil has in some countries been banned from being used in packaging as it is deemed to have too high an environmental cost. As the vacuum deposited aluminium coatings are often around one hundredth of the thickness of the rolled foils these have been targeted at replacing many of the packaging foil products (1). This foil replacement application is one of the highest growth markets.

When these coatings are examined in detail it becomes apparent that they are not perfect but contain a large number of defects. A detailed examination of the supply materials, previous processing and the vacuum deposition process show that there are many factors that can affect the integrity of the coatings which in turn affect the resultant barrier performance.

Even with these less than perfect coatings the market for the vacuum deposited barrier films is huge with approximately 550,000 tonnes of vacuum coated products being sold into the packaging industry annually and a predicted growth of ~5% per annum. This represents the coating of approximately 22,000 million square metres of material. Of these packaging materials metallised polypropylene takes the largest share at over 50% with metallised PET being the second most widely used substrate. The market continues to grow partly encouraged by environmental pressures with metallised polymers being used to replace aluminium foil and also replace tin cans. Within the area of vacuum deposited coatings there is a difference in market growth expectation for different materials. The deposition of metals, primarily aluminium, has been done for more than 50 years whereas the deposition of the transparent barrier materials is relatively new. It is only relatively recently that the costs have reduced enough, as well as the banning of a chlorine containing coating, to make the transparent barrier vacuum deposited coatings attractive to the packaging industry. This market sector of transparent barrier coatings has been growing at in excess of 20% although from such a small volume this still makes the volumes small by comparison to the metallised films.

When we look at different barrier coatings we can group them into specific types such as packaging, intermediate and ultra-barrier coatings and then subdivide these into opaque or transparent barrier materials.

1.1 Packaging

Packaging has to achieve a number of different functions. Ideally it provides containment to keep the product secure. It has to be convenient to use providing an opportunity for communication, have suitable aesthetics, be non-toxic, tamper-resistant (or tamper-evident), be functional in size & shape and compatible with the production process and the product it contains, low cost, recyclable, reusable or disposed of easily. In addition it has to preserve the product by providing protection against environmental (oxygen, water/moisture, light, chemical attack, contamination from micro organisms), physical attack (such as rodents, and insects), and mechanical hazards (handling damage) during storage and distribution. So when incorporating a barrier coating it needs to be complimentary to the existing substrate properties.

The largest volume of vacuum deposited packaging materials is used for the packaging of food. Often this market segment is driven by minimising cost and as vacuum coating adds cost over the basic flexible webs there has to be a cost benefit to justify using this coating process.

Extending the shelf-life of products is one of the most easily proven cost benefits that can used to justify the addition of vacuum deposited coatings. If we take an example of potato crisps/chips if we open the pack and the crisps are left in the open then moisture will be absorbed and the crisps become soft and soggy. If they were left out in the air and in daylight over a period of time the taste of the crisps would decline as the fats turn rancid because of degradation by oxidation by the oxygen in air or photo-oxidation by daylight. The same pack of crisps can be left for weeks on the shelf and when opened will still be crisp and with the same taste as when first made because the vacuum deposited coating has provided a barrier to the oxygen, moisture and light keeping the crisps dry and fresh. Providing this superior barrier performance means that the bags of crisps do not have to be sold within a few days of manufacture but can still be safely sold weeks later and so the waste and loss of profit is reduced.

The manufacturer of any food product will know what quantity of moisture, or oxygen or light will cause the product to degrade. The manufacturer will also choose a shelf-life that they wish to achieve and this information can be used to calculate how good the barrier performance of the packaging has to be to achieve these goals.

Most of the time we think of the barrier being to prevent things getting into the food but the reality is that it also prevents things escaping from the food too. If we think of water vapour it can turn food soggy but if lost from food it can allow the food to dry out too much. The drying out of food can be a problem for foods such as breads and cakes. A less obvious problem is in frozen food where the loss of moisture through the sublimation of ice can lead to freezer burns of the food.

Oxygen from the air can oxidize some materials such as fats turning them rancid but also can oxidize vitamins such as vitamin C, causing it to lose potency. However oxygen is not the only gas that can be controlled by barrier coatings and this is used to advantage in controlled or modified atmosphere packaging (MAP). In modified atmosphere packaging the package is flushed out using a gas, such as dry nitrogen, and then the package is filled with a specific gas or mixture of gases. In this case the barrier is designed not only to keep the air out but also to keep the modified gas composition inside the package. The gas being used to fill the package might be used to slow down the ripening of fruit and so extending the shelf-life or it may be used to maintain the colour of the food which can be more about aesthetics than food safety.

Light can be quite detrimental to food with photocatalytic reactions causing the degradation of fats, flavours, vitamins, such as vitamins A, B12, D, E, K, etc and changes of colour. So one early choice is to decide if the product needs to be protected from light and so have a metal coating deposited as the barrier or if light is not a problem then it may be preferable for the foods to be visible to the customer and these would have a transparent barrier coating deposited.

The packaging also needs to be benign and not interact with the product either. The polymer may absorb aromas from the foodstuffs and this may reduce the aroma detected by the consumer. This process of aroma absorption is known as scalping. The packaging also should not taint the foodstuffs by losing anything from the polymer into the foodstuffs known as migration.

1.1.1 Opaque Barrier

Opaque light barrier vacuum deposited coatings are achieved primarily by aluminium metallization. The opacity of the very thin metal coating is usually quoted as the optical density of the coating. Opacity is a measure of light incident on the coating divided by the amount of light transmitted through the coating. The optical density (OD) of a coating is the opacity expressed as a logarithm to base ten. This measurement uses a white light source and detector. The transmitted light value can be obtained before deposition starts each time to establish the 100% value and so the substrate is eliminated from the measurement.

$$\text{Opacity} = \frac{\text{Incident light}}{\text{Transmitted light}}$$

$$\text{Transmittance} = \frac{\text{Transmitted light}}{\text{Incident light}}$$

$$\text{Opacity} = \frac{1}{\text{T}}$$

$$\text{Opacity Density} = \log_{10} \text{Opacity} = \log_{10} \frac{1}{\text{T}}$$

One packaging company has a requirement of a shelf life of 49 days for packaging their potato chips. Light will turn the chips rancid in only 3 days and so they require an opaque thin metal coating for which they have assessed that an OD of 1.7 will achieve the 49 days. Another customer requiring a 90 day shelf life needs a thicker coating to block out more light and an OD of more than 2.2 is necessary (2,3). At the same time as the light needs to be blocked the moisture ingress needs to be limited too. The moisture content of the chips after processing is between 1.3%–1.8% and the acceptable limit at the end of the 49 day shelf life is 2.5%. This means that an increase of only 0.7% can be allowed in 49 Days. This increase can be converted into a weight increase and can provide the target for the acceptable water vapour barrier performance. We can look at this type of calculation in Chapter 4 on materials. In their case the metallized oriented polyester met their requirements but was too expensive and polypropylene met most of their requirements. It was found that improving the surface smoothness of the polypropylene film improved the performance enough to achieve all their requirements and at a lower cost than metallized polyester.

Examination of the metallized film has shown that not only do defects affect the barrier performance of the metallized polymer film but also that the handling of the film in the downstream processes, such as laminating or filling, will increase the number of defects and further degrade the barrier performance. Hence when designing and calculating the barrier packaging there needs to be some allowance for this reduction in performance during packaging processing (4,5).

The opaque packaging is primarily done using aluminium metal and the cheapest metallization process for this is by resistance heated evaporation sources. This technology is mature and the vacuum metallizers have been developed over the years such they can now be built to 4.45m width and with a maximum winding speed of 1250 m/min.

1.1.2 Transparent barrier

Transparent barrier packaging becomes essential where it is desired that the product being packaged is seen by the consumer or user (6,7). Transparent barrier has the added benefit that it makes online metal detection easier. Also electronic devices such as OLEDs where there is a display that needs to be read by the consumer or a photovoltaic device where the light needs to pass through the barrier material to reach the photovoltaic device to be converted into electricity the barrier materials also need to be transparent. The most widely used transparent materials have been alumina and silica (8,9). Both of these materials have been deposited by a variety of different means all aimed at trying to reduce the costs. Over the last ten years the costs have fallen considerably but the production of the transparent barrier materials still remains at a cost of at least twice that of aluminium metallization. Silica has been deposited by thermal evaporation, electron beam evaporation, induction heating evaporation and chemical vapour deposition. Alumina has been deposited by electron beam reactive evaporation until recently when a number of companies have produced material using modified resistance heated evaporation metallizers (10–13). The aim of this work was to take a standard metallizer and with a slight modification convert the opaque aluminium into a transparent alumina whilst maintaining as much of the original winding speed as possible. In this way it was hoped to bring the costs down to a similar level of aluminium metallizing.

There have been occasional other coatings developed (14,15) that have been championed by the company or institute that developed them but as yet none have shown either sufficient cost or technical advantage to make the technology be taken up very widely. An example of this would be the evaporation of melamine barrier coatings (16,17).

The need for transparent ultra barrier coatings has caused the whole barrier technology to be reviewed and developed in order to improve the standard packaging grade barrier coatings by several orders of magnitude. Not only was the quality of the substrate surface improved but also different inorganic coatings were investigated. Until the problem with pinholes was understood and improved there was no need to test if the inorganic coatings were the best for the job. Once improved surfaces had been produced, enabling more defect free vacuum deposited coatings to be deposited (18–20), different inorganic coatings such as indium tin oxide, silicon nitride and carbon or hydrogenated versions of silicon nitride were used to produce ultra barrier materials for evaluation (21–26). This area of development has not yet been completed and I would expect that there will be many more inorganic materials evaluated either as individual layers or in combination with other inorganic layers either as discrete or merged layers. There is the thought that to improve the water barrier performance modifying the chemical composition to make the coating hydrophobic might have some advantages and might be achieved by adding fluorine (27,28). The concern over making the surface hydrophobic is that it would also make the surface harder on which to stick other layers, either as additional vacuum deposited layers or by lamination.

The evaporation process tends to be very fast but the structure of the coating can contain defects relating to the nucleation and growth. The electronics market can, at least in the development phase, withstand a higher price for producing the ultra barrier materials. A number of groups have used sputtering as the deposition process even though these sources may deposit coatings much more slowly, even as much as three orders of magnitude more slowly. The advantage of the sputtered coatings can be the production of higher density coatings with fewer morphological defects. Work is being done to use additional plasma along with evaporation sources to match this densification but with the higher deposition rates.

One additional difficulty in the deposition of some of these ultra barrier coatings for electronic applications, that is not present for the food packaging applications, is the need to deposit the coatings over surface features. In photovoltaic cells there are several laser cut trenches made through various layers to connect or separate individual cells. Thus the polymer layers that are used as separation layers in multilayer transparent barrier coatings have to be conformal. It is possible to smooth out some of these features by depositing a thicker initial polymer layer first (29).

1.2 Markets

The information on markets can appear misleading because some of it is described in area of coated material others quote weight of material coated and still others will quote value. As the price of aluminium coated material for packaging can be a hundred times cheaper than a transparent ultra barrier material and the substrate thickness of the ultra barrier may be ten times as thick for the electronics applications compared to the packaging barrier materials the proportions can be skewed in different ways.

The largest area of material to be coated is for the opaque packaging market and is from simple aluminium metallizing machines. This market is mature and still growing at around 5% per annum. The amount of growth depends on where in the world you are with India and China growing substantially more than the more mature European and USA markets. The Asia pacific region has grown very rapidly and from supplying <20% capacity of metallized films in 2002 it now is the largest supplier with >33% world capacity.

The markets that are currently grabbing the headlines are the transparent barrier packaging markets as these are growing at between 10%–15% per annum worldwide and more than 20% in Europe which is considered a mature packaging market and so for most barrier materials is lower than average. Of these the largest growth is in the retortable transparent barrier materials.

With the interest in reducing energy use there is renewed interest in reducing materials use and energy used in materials production and energy used in transportation. Tin cans are steadily being replaced by stand-up pouches. This segment of the market for metallized film has been growing quickly but the growth of the transparent barrier retortable materials for this market is growing at more than double the rate of the metallized materials. Again there is different interest in the different world regions with approximately half of the vacuum coated films in Japan being transparent barrier films. This transparent market got a boost when Japan banned polyvinyl dichloride (PVdC) and the vacuum coated transparent barrier was one of the few materials able to replace it. Since then Europe has followed this course of action and growth of transparent vacuum deposited barrier coatings has increased similarly.

The markets that are quoted as having the largest potential growth are the ultra barrier materials as these are linked to two huge markets, the display industry with the OLEDs and the flexible photovoltaic market. To realise this potential market the performance, reliability and reproducibility of these ultra barrier coatings has to be proven. Also there is an expectation that as the deposition process is developed and production scaled up the price will decline. Currently, although there have been announcements of material becoming available it is hard to get rolls of material, many offering only A4 sheets for evaluation.

To put some numbers on these markets is difficult. The world market for vacuum deposited products is estimated to be of the order of 800 kT for 2009 which is approximately 32,000 Million sq.m. of coated area. However this figure tends to neglect the specialist coatings markets as they are fragmented and small by comparison to the packaging markets that dominate because of the quantity of materials coated and as the substrates tend to be thin with much of the materials coated being around 12 microns thickness this results in the area coated being similarly huge.

If we look at the photovoltaic market for barrier coatings it is currently negligible but once products are widely available this will grow very rapidly. What this means in terms of tonnes of material or area of material coated is more of an unknown. Photovoltaics are predicted to grow at more than 20% per annum for some time to come. Flexible photovoltaics are predicted to grow faster than this as they take some market share. The growth is often quoted as an increase in gigawatts of installed conversion capacity such as growth from approximately 6GW in 2008 to 35GW by 2015. The area of this increase in installations will depend on the efficiency conversion of the type of cells used. This can vary from less than 5% to close to 20% for standard cells or arrays. So calculating the area from the energy can vary widely. As flexible cells can be deposited on metal foil substrates or polymer substrates the need for barrier coatings can either be for one side barrier only in the case of metal foils or both sides for the polymer substrate materials. This again potentially adds a large error to any estimates of the requirements barrier coatings. In reality it probably does not matter what the numbers are other than it is likely to be growing to require several million square meters over the next few years. As the profit margins for the ultra barrier materials are expected to be larger than for food packaging applications this makes this type of barrier coating a very attractive target.

A similarly humungous market growth is forecast for OLEDs by the likes of IDTechEx and Pira International with an estimated 35% per annum growth over the next five years taking the market from an estimated $615Million to a more than ten fold increase. This could represent 500–750 million display devices per year. Again the guess of the required area of protective barrier materials is difficult as this type of figure can include devices that have more than 100 small displays per square meter to much larger displays for computer screens, etc. However it is probably safe to believe that this too will require a few million square meters of ultra barrier coatings. This requirement for ultra barrier coatings looks to be required for many different technologies as a number of the newer photovoltaic and display devices are all sensitive to moisture or oxygen or both. Thus even if the preferred product technology changes the ultra barrier for encapsulation will still be required.

Between the packaging barrier products and the ultra barrier materials for the electronics markets is the market for materials with intermediate barrier performance needed by the vacuum insulation panel markets (30–32). This is where an insulation material is encapsulated by a vacuum coated barrier material. The insulation material is evacuated before sealing and the panel acts something like a vacuum flask and can be used in white goods or buildings as an insulation material. The requirement for the materials to be included in buildings is that the performance will still be good in anything from 50 years to 100 years time. This is usually achieved by having a barrier performance half way between food packaging barrier and the ultra barriers needed for the sensitive electronic materials and then assisting the barrier by adding a scavenger material embedded in the insulation to absorb a quantity of gas or moisture that does pass through during the 50 or more years. An alternative design has used a gas filled insulation panel with the barrier preventing the exchange of gas from the inside and being replaced by gas from the outside (33).

Although this market is growing there are few predictions about how big it will become. The rate of building varies enormously and there are many other competing materials that can be used to deliver an equivalent performance and it is not clear which materials will become favoured. These materials are not required to be transparent and so laminated metallized materials can be used. Using double side and laminations of multiple metallized films can be used to achieve the desired level of permeability. Thus the growth of this market is probably hidden within the continued growth of metallized films.

Of all these dispirit markets the need for barrier for electronic applications would appear to have a difficult technical specification but the market looks to be stable and growing well for some time to come.

At the other end of the spectrum the packaging market that is currently by far the largest market in general has the lowest margins with a more easily achieved technical target but is also the most fickle of the markets. There are the conflicting requirements of preserving food for longer but also reducing the quantity of packaging and the need to increase re-cycling. This, in many respects, makes this the harder market to participate in.

References

1. Decker W. et al. 'Metallized polymer films as replacement for aluminium foil in packaging applications' 47th Ann. Tech. Conf. Proc. Society of Vacuum Coaters 2004, pp 494–499.
2. Gavitt I.F. 'Vacuum coating applications for snack food packaging' 36th Ann. Tech. Conf. Proc. Society of Vacuum Coaters 1993, pp 254–258.
3. Gavitt I.F. 'Vacuum coating applications for snack food packaging' 37th Ann. Tech. Conf. Proc. Society of Vacuum Coaters 1994, pp 127–132
4. Specht J. 'Metalization: An End-User's Perspective' 41st Ann. Tech. Conf. Proc. Society of Vacuum Coaters 1998, pp 440–445.
5. Hoekstra T. 'Metallized materials for FMCG packaging trends and needs' Proc. 3rd Ann. Vac. Coating & Metallizing Conf. 2004, Section II
6. Brody A.L. 'Glass-coated flexible films for packaging: an overview.' Packag. Technol. Eng. Feb 1994 pp 44.
7. Dodrill D. 'Advances in clear high barrier packaging' Medical Device Manufacturing and Technology 2006, pp 45–48.
8. Seserko P. 'Transparent barrier coatings and the EB method' Paper Film and Foil Converters Feb 1999, pp 62–65.
9. Chahroudi D. 'Transparent glass barrier coatings for flexible film packaging' 34th Ann. Tech. Conf. Proc. Society of Vacuum Coaters 1991, pp 130–133.

10. Schiller S. et al. 'How to produce Al_2O_3 coatings on plastic films' Proc. 7th Internat. Conf. Vacuum Web Coating. 1993, pp 194–219.
11. Kelly R.S.A. 'Development of clear barrier films in Europe' 36th Ann. Tech. Conf. Proc. Society of Vacuum Coaters 1993, pp 312–316.
12. Kelly R.S.A. 'Development of aluminium oxide clear barrier films' 37th Ann. Tech. Conf. Proc. Society of Vacuum Coaters 1994, pp 144–148.
13. Kobayashia T. 'Reactive vacuum vapor deposition of aluminum oxide thin films by an air-to-air metallizer' *J. Vac. Sci. Technol.* A 24,4, Jul/Aug 2006, pp 935–945.
14. Moser E.M. et al. 'Hydrocarbon films inhibit oxygen permeation through plastic packaging material' *Thin Solid Films* Vol. 317, Issues 1–2, 1 Apl 1998, pp 388–392.
15. Aisenberg S. 'Diamond-like carbon deposition technology for improved barrier films' Proc. 4th Internatl. Conf. Vac. Web Coating. 1990, pp 25–38.
16. Jahromi S. 'A new development in clear barrier coatings' 51st Ann. Tech. Conf. Proc. Society of Vacuum Coaters 2008, pp 799–802.
17. Jahromi S. and Moosheimer U. 'Oxygen barrier coatings based on supramolecular assembly of melamine' Macromolecules 2000, 33, pp 7582–7587.
18. Yailizis A. et al. 'Ultra high barrier films' 43rd Ann. Tech. Conf. Proc. Society of Vacuum Coaters 2000, pp 404–407.
19. Graff G.L. et al. 'Fabrication of OLED devices on engineered plastic substrates' 43rd Ann. Tech. Conf. Proc. Society of Vacuum Coaters 2000, pp 397–400.
20. Weaver M.S. et al. 'Flexible Organic LED Displays' 44th Ann. Tech. Conf. Proc. Society of Vacuum Coaters 2001, pp 155–159.
21. Assche, F.J.H.v., et al. 'A thin film encapsulation stack for PLED and OLED displays' Soc. Info. Display (SID) Internatnl. Symposium, 2004. 35: pp 695–697.
22. Erlat A.G. et al. 'Mechanism of water vapour transport through PET/AlOxNy gas barrier films' *J. Phys. Chem. B*, 108, 2004, pp 883–890.
23. Erlat A.G et al. 'Ultra-high barrier coatings for flexible organic electronics' 48th Ann. Tech. Conf. Proc. Society of Vacuum Coaters 2005, pp 116–120.
24. Henry B.M. et al. 'Microstructural and gas barrier properties of transparent aluminium oxide and indium tin oxide films' 43rd Ann. Tech. Conf. Proc. Society of Vacuum Coaters 2000, pp 373–378.
25. Fahlteich J. et al. 'Mechanical and barrier properties of thin oxide films on flexible polymer substrates' 51st Ann. Tech. Conf. Proc. Society of Vacuum Coaters 2008, pp 808–813.
26. Miyamoto T. et al. 'Gas barrier performances of organic-inorganic multilayered films' 44th Ann. Tech. Conf. Proc. Society of Vacuum Coaters 2001, pp 166–171.

27. Teshima K. et al. 'Gas barrier performance of surface-modified silica films: dependencies on surface functional groups of the films' 47th Ann. Tech. Conf. Proc. Society of Vacuum Coaters 2004, pp 691–692.
28. Hozumi A. at al. ' Fluoroalkylsilane monolayers formed by chemical vapour surface modification on hydroxylated oxide surfaces' Langmuir: the ACS *J of surfaces and colloids,* vol.15 n.22, 1999, pp 7600–7604.
29. Gross M.E. et al. 'Ultrabarrier protective coatings for atmospherically sensitive thin-film electronic devices' 46th Ann. Tech. Conf. Proc. Society of Vacuum Coaters 2003, pp 89–92.
30. Tenpierik M. and Cauberg H. 'Vacuum Insulation Panel: friend or foe?' PLEA2006 – Proc. 23rd Conf. on Passive and Low Energy Architecture, Geneva, Switzerland, 6-8 September 2006.
31. Binz A. et al. 'Vacuum Insulation in the Building Sector - Systems and Applications' HiPTI - High Perf. Thermal Insulation - IEA/ECBCS Annex 39, Final Report 2005 pp 1–111.
32. Musgrave D.S. 'Finite element analysis used to model VIP barrier film performance' Proc. 7th Internatl. Vac. Insulation Symp. Sept. 2005.
33. Griffith B. and Aresteh D. 'Gas filled panels: A thermally improved building insulation' Proc. ASHRAE/DOE/BTECC Conf. Thermal performance of the exterior envelopes of buildings V. Dec. 1992.

2

Terminology

If we imagine we have a product encased in a barrier package where we have either a vacuum or inert gas within the package and then we examine a section of the package. On the product side of the barrier material we have one pressure and concentration of each gas or water vapour and this will be different to the pressure and concentration of the same gases or water vapour outside the package. This difference provides the driving force for the transfer of the gases or water vapour from one side to the other in order that the system comes to equilibrium.

To achieve equilibrium there are some key events that occur for all barrier materials such as absorption or sorption, diffusion and desorption. On the high concentration side of the barrier material the gas is is absorbed (or sorbed) and it then diffuses through the bulk material until it reaches the opposite surface where it is then desorbed so raising the concentration inside the package. This process gives us the following terms.

Sorption (at the interface)
Gases, vapour or dissolved chemicals or suspended substances are adsorbed at the surface of the solid.

Desorption
The adsorbate leaves the solid as a gas.

The absorption and desorption of gas is defined by Henry's Law where Henry's Law states that at a given temperature the amount of gas dissolved in a material is proportional to the pressure of gas above it. This is expressed in the equation shown in Fig. 2.1. Henry's Law only applies where there is no chemical reaction between the gas and the material. An example where there can be a surface chemical reaction is the reaction of oxygen with an aluminium coating where the oxygen will not necessarily diffuse through but may instead convert the aluminium to aluminium oxide.

The sorption is not only important as part of the gas transport through the barrier material but is also important as part of the product contact stability evaluation. If we use food packaging as our example it is possible for sorption to occur of components of the food such as aromas, flavours or other food components (1–7). Sorption of flavours or aromas is also known as scalping. This can degrade the food and can give rise to loss of flavour or aroma as well as providing a source of taint if the packaging is reused. The complimentary contamination process is where there is desorption of something from the package into the food which is known as migration (8–10). This can be something that is inherent in the polymer such as the oligomer or additives used in processing or something remaining from earlier sorption in the case of reusable packaging.

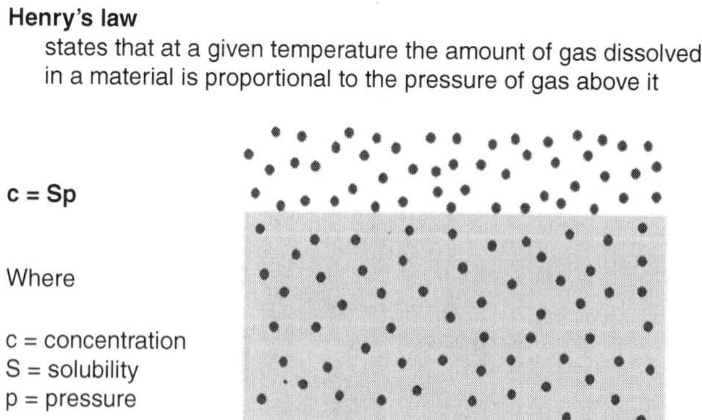

Figure 2.1 Henry's Law relating the pressure and concentration of gas to solubility.

The sorption has many factors that affect the ease of absorption such as the shape, size, concentration and polarity of the material that is absorbed. The polarity, orientation, crystallinity, cohesive energy density, packing and glass transition temperature of the polymer will also affect the speed and amount of absorption as too do factors such as temperature, time, acidity/alkalinity and if there is any co-sorbate. The general rule of thumb is that the smaller the size, the higher the concentration or higher the temperature the faster the absorption and the greater the amount to be absorbed (1).

Diffusion

The transport of matter from one point to another by random molecular motions. It occurs in gases, liquids, and solids. Where diffusion through a solid is where the permeate penetrates the solid material through pores or molecular gaps.

So if we look at our barrier material and assuming we are using it for an oxygen barrier if we look at Fig. 2.2 on the right had side we have the random motion of gas atoms zipping about and colliding with each other and with surfaces. If the concentration and pressure are higher than in the barrier material the oxygen colliding and sticking to the surface will move into the barrier material and diffuse through. The rate of diffusion will depend on the materials used for the barrier. Diffusion will occur in all directions in the material but

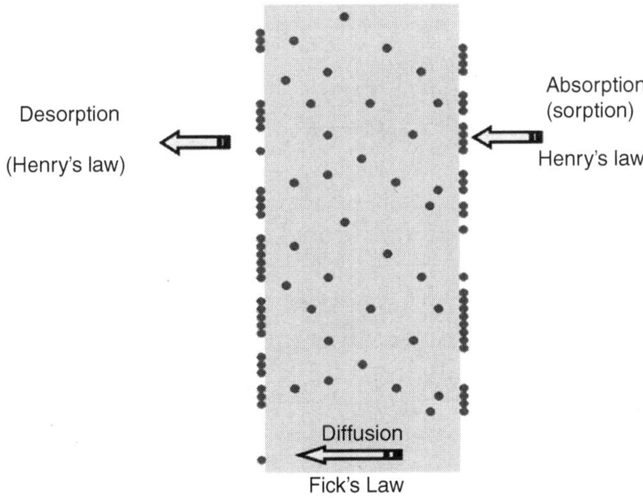

Figure 2.2 A schematic of the basic mechanism of a barrier material.

as there is a pressure difference the driving force for the diffusion is the difference in gas concentration between the inside and outside this will give the diffusing gas a direction. Once the gas reaches the inner surface it will want to desorb and once it does this it will then form a part of the gas within the package and will continue with the random motion within the gas contained in the package.

As with the sorption many of the same factors will affect the diffusion however the effect may not be at the same rate as for the sorption.

The combined process of absorption, diffusion and desorption gives us the permeability.

Permeability

The property or condition of being permeable. The rate of flow of a liquid or gas through a porous material. How much penetrates in a specific time, dependent on type of permeate, pressure, temperature, thickness of the solid & the area.

Permeation

Permeation, in physics, is the penetration of a substance (permeate) through a solid.

where

Permeation = Permeability/film thickness

Permeate

The substance permeating through the solid

Permeance (not commonly used)

A measure of a material's resistance to water-vapor transmission, expressed in perms.

Equal to the ratio of (a) the rate of water vapor transmission through a material or assembly between its two parallel surfaces to (b) the vapor pressure differential between the surfaces.

We want to be able to measure the permeability of barrier materials and we have seen in Fig. 2.1 we have Henry's Law for the absorbtion and desorbtion. If we look at Fig. 2.3 we have Fick's 1st law of diffusion which allows us to calculate the diffusion rate. From these two calculations we can find the permeability.

P = Permeability coefficient $\quad\quad$ P = (D)(S)

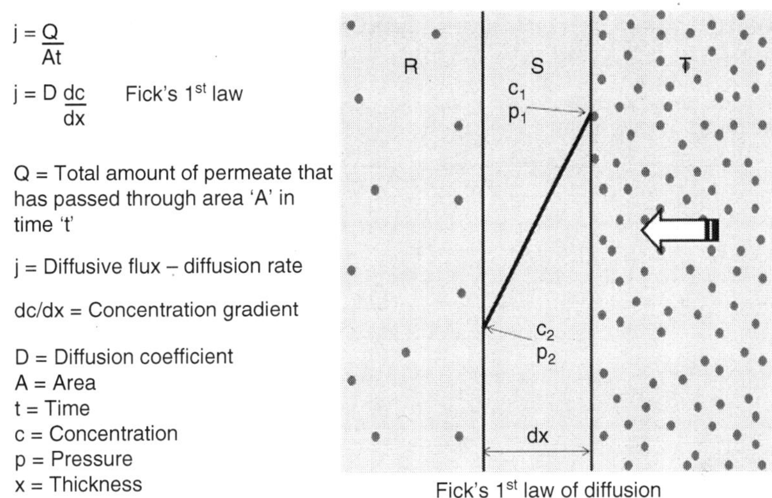

Figure 2.3 Fick's first law of diffusion.

where

D = Diffusion coefficient which is a measure of how rapidly penetrant molecules are moving through the barrier, in the direction of lower concentration or partial pressure. D is a kinetic term that describes how fast molecules move in a polymer matrix.

S = Solubility coefficient which is the amount of transferring molecules retained or dissolved in the film at equilibrium conditions. S is a thermodynamic term that relates to how many molecules dissolve in a polymer matrix.

The diffusion coefficient for materials is an important measure and it helps to explain why barrier materials can differ in performance and how some of the structures are modified to compensate for the coating limitations. Diffusion of gases takes place in all states so for our oxygen there will be a diffusion coefficient for oxygen in air which will be different to the diffusion coefficient for oxygen in water or for oxygen in a solid. The diffusion coefficient will also differ for different materials and this is shown in Table 2.1.

Some things are worth noting such as the values for the polymers. The polymer films may be cast films or uniaxially or biaxially oriented and the resultant films are different. During polymerisation and subsequent processing the degree of crystallinity can be controlled. Crystalline polymers are better barriers and have a lower

Table 2.1 The diffusion coefficient for oxygen is some different materials.

e.g. Oxygen in air	$D_{O_2} = 0.15\ cm^2/s$
e.g. Gas in a liquid	$D_{O_2} = 0.0000197\ cm^2/s$ @ 20°C $\cong 2 \times 10^{-5}\ cm^2/s$
e.g. Oxygen in solid	
Oxygen in Quartz	$D_{O_2} = 10^{-25}\ cm^2/s$
Oxygen in Silica	$D_{O_2} = 10^{-13}\ cm^2/s$
Oxygen in Polypropylene	$D_{O_2} = 10^{-7}\ cm^2/s$
Oxygen in Polyester	$D_{O_2} = 10^{-9}\ cm^2/s$

diffusion coefficient to amorphous polymers of the same chemical composition. Hence different manufacturers or different grades of the same polymer material may have very different diffusion rates depending on the processing.

In looking at Table 2.1 as a whole what is important to note is the magnitude of the diffusion coefficients. If we look at the diffusion coefficient for oxygen in air and oxygen in silica we can see that there is twelve orders of magnitude difference between them. So if we consider a silica coated polymer film for our barrier material and we know we will have some pinholes in our silica coating we can see why our barrier performance falls short of the ideal performance. Wherever there is a pinhole instead of silica we have air with a diffusion coefficient twelve orders of magnitude higher than that of silica. Hence air or water vapour will be rushing through the pinholes and this is why these pinhole defects dominate the barrier performance and why if we can eliminate them the barrier performance improves so much. Often we cannot eliminate the pinholes and so if we go back to Table 2.1 and look at the diffusion coefficient for polymers we can see that they are somewhere between air and silica being six orders of magnitude better than air but also a few orders of magnitude short of the performance of silica. Thus if we overcoat our ceramic or glasslike barrier coating with a polymer that can fill the pinholes we can significantly reduce the diffusion coefficient in the defect pinholes and so improve the overall barrier performance. If this process is taken further and the polymer overcoating is filled

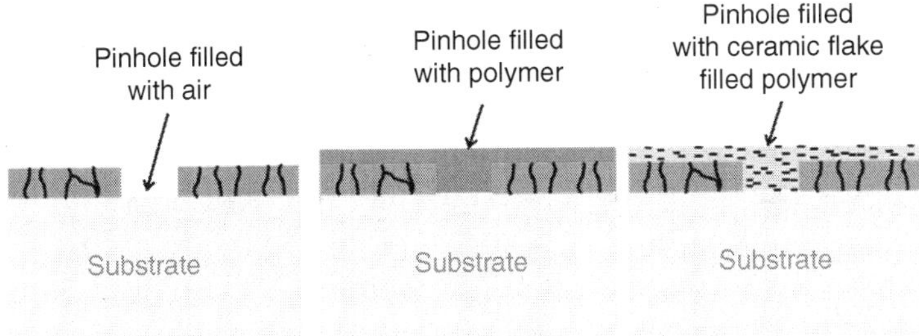

Figure 2.4 A schematic changing the medium filling a pinhole.

with glasslike or ceramic flake materials the diffusion coefficent can be raised further and become much closer to the rest of the silica barrier coating. This is shown schematically in Fig. 2.4.

The diffusion coefficients only apply where the gas or water vapour is dissolved into the matrix and then can diffuse through. In the vacuum deposited coatings there will be a number of defects such as pinholes, as described above, pores and grain boundaries. The pores can be where grains have not yet met and so leave a small hole through the material. Depending on the size of these holes will depend on the rate of diffusion through them. If they are gross defects and are large enough, such as pinholes would be, the diffusion will be close to that of diffusion through air.

Below the level of the pinholes the defects can be graded by size and the mechanism for the gas to pass through changes with size. This is shown in Fig. 2.5. The decreasing size and change to each different mechanism corresponds with a decreasing amount of gas passing through the defects. Overall the barrier performance for any coating will depend on the number and size of the pinholes and defects and the proportion of the total surface area that these represent. Water vapour may have different mechanisms and it has been suggested that water may be attracted to, or interacts with, the surfaces, due to the polar or reactive nature of the water molecule.

The difference in diffusion between packaging polymers such as polypropylene or polyester and silica also shows why the thin vacuum deposited barrier coating so dominates the barrier performance of the packaging barrier materials even where there are several layers laminated together.

The measure that is usually used to compare the performance of the different barrier materials is the permeability (12–27). To find

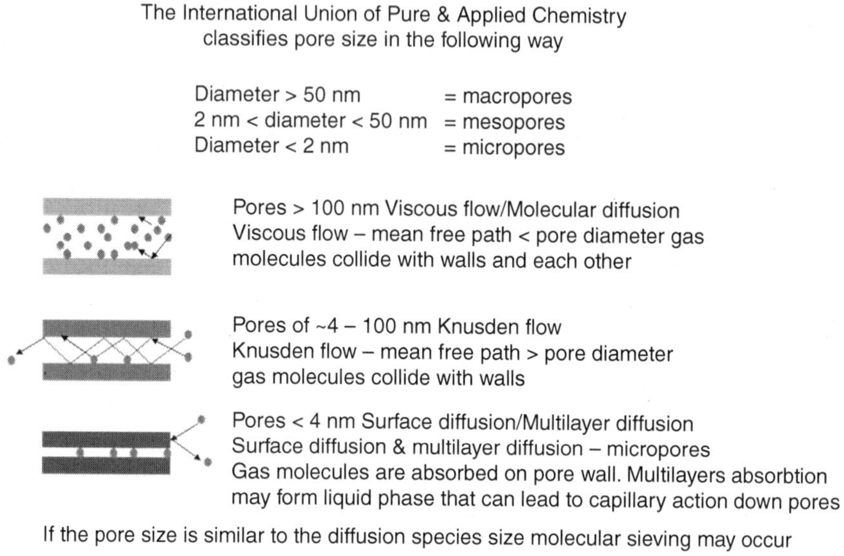

The International Union of Pure & Applied Chemistry
classifies pore size in the following way

Diameter > 50 nm	= macropores
2 nm < diameter < 50 nm	= mesopores
Diameter < 2 nm	= micropores

Pores > 100 nm Viscous flow/Molecular diffusion
Viscous flow – mean free path < pore diameter gas molecules collide with walls and each other

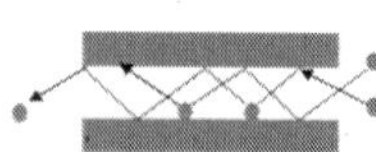

Pores of ~4 – 100 nm Knusden flow
Knusden flow – mean free path > pore diameter gas molecules collide with walls

Pores < 4 nm Surface diffusion/Multilayer diffusion
Surface diffusion & multilayer diffusion – micropores
Gas molecules are absorbed on pore wall. Multilayers absorbtion may form liquid phase that can lead to capillary action down pores

If the pore size is similar to the diffusion species size molecular sieving may occur

Figure 2.5 The classification of defect size and the resultant gas movement process (11).

the permeability if we go back to Figs. 2.1 & 2.3. The rate of diffusion 'j' of matter 'i' is described by Fick's 1st law, which states:

$$j_i = D_{ij} \frac{dc_i}{dx}$$

where Dij is the diffusion coefficient of the matter 'i' in a medium 'j', and describes the concentration gradient in the diffusion direction.

The solubility of a gas in solid materials is given by Henry's Law which states:

$$c_x = S.p_x$$

Now if we substitute Henry's law into Fick's 1st Law we can obtain the molar flux:

$$j_i = \frac{D_{ij} \cdot S \cdot (p_1 - p_2)}{x} = P^*_m \cdot \frac{Dp}{x}$$

where

c = concentration
p = pressure
x = thickness

Table 2.2 Some example values of material permeation and permeability.

Polymer	Permeability $\left[\frac{cm^3 O_2}{md\ atm}\right]$	Film Thickness [μm]	Permeation $\left[\frac{cm^3 O_2}{m^2 d\ atm}\right]$
Low density Polyethylene	$3.61e^{-1}$	60	6000
High density Polyethylene	$6.70e^{-3}$	60	100
Polypropylene	$1.51e^{-1}$	60	2500
Polyethylene Teraphthalate	$1.20e^{-3}$	12	100
Permeation = Permeability/film thickness			

and $P_m^* = D_{ij} \cdot S$ is the molar Permeability of the film normalized to a unit thickness. Dividing P_m* by the film thickness 'x' delivers the Permeation P:

$$P = \frac{P_m^* \cdot M}{p \cdot x} = \text{permeation in cc/sq m.day.atm}$$

How the relationship between permeation and permeability works in practice is shown for a few materials in Table 2.2

When either making measurements of permeability or comparing values it is very important to establish that the parameters used are compatible and that the comparison is like with like. The pressure difference between the two sides of the barrier will affect the driving force, the larger the area of the barrier material the greater the value will be, the thicker the barrier material the lower the value will be and the time needs to be normalised as the longer the time the greater the quantity of gas that will have passed through the barrier material. Some materials are affected by humidity and so this too needs to be standardised as too does the temperature.

For diffusion of gases through solids, permeability P = g/cm/s/atm is the product of the diffusivity D = cm^3/s and the solubility S = $g/cm^3/atm$. The water vapour transmission rate (WVTR) is commonly reported in units of $g/m^2/day$ but must be qualified by including the temperature and relative humidity (%RH).

There are occasionally ratios given of the permeability of different gases so that if the permeability of one gas is known then the others

Table 2.3 The gas transmission rates of various gases at two different temperatures through PET film normalised to oxygen transmission rate.

PET film 20 microns	**Helium**	**Nitrogen**	**Air**	**Oxygen**	**Carbon dioxide**
Room temperature	48.20	0.18	0.39	1	6.23
40°C	36.00	0.17	0.33	1	4.94

Table 2.4 The gas transmission rates of various gases through different films at room temperature normalised to the oxygen transmission rate.

Film type	**Helium**	**Nitrogen**	**Air**	**Oxygen**	**Carbon dioxide**
PET 20 microns	48.20	0.18	0.39	1	6.23
PC 125 microns	9.17	0.21	0.40	1	4.54
PVdc 30 microns	30.89	0.12	0.23	1	3.47

Table 2.5 A comparison of effective diameter of various gases.

	Helium	**Nitrogen**	**Air**	**Oxygen**	**Carbon dioxide**
Molecular Weight	4	28	29	32	44
Dynamic Diameter nm	0.26	0.364	0.34	0.346	0.33

can be estimated (28). Tables 2.3 and 2.4 show that the ratio between gases differs for different materials and different temperatures and so these ratios need to be used with caution.

If we compare the dynamic diameters of the different gases in Table 2.5 we can see that oxygen and carbon dioxide are similar in size but if we look at the transmission values in Tables 2.3 and 2.4 we can see that they are very different. This is because the carbon dioxide has a higher critical temperature for the polymer and hence the largest solubility coefficient giving rise to the largest permeation value. Nitrogen and helium have similar solubility coefficients and so the permeation difference relates to the difference in effective diameters with the smaller diameter finding it easier to diffuse through the polymer. As the reason why the permeability varies for different gases it is not possible to use any factor to correct the ratios easily. It has been suggested that errors are typically more than 20% for data from the same source and much more than this for data from varied sources (29).

Most of the time the final barrier material is a structure of many different components such as polymer substrate, vacuum deposited coating, adhesive, laminate and possible multiple adhesives and laminates. As we want to know the final barrier material permeation performance we need to combine the permeation performance of each component layer.

The calculation for the total permeability can be expressed using the ideal laminate theory (ILT) as follows;

$$\frac{1}{P_{total}} = \frac{x_p/x}{p_p} + \frac{x_c/x}{p_c} + \frac{x_l/x}{p_l} + \ldots$$

where

x = total multilayer material/laminate thickness
x_p = polymer film thickness
x_c = vacuum coating thickness
x_l = laminate or other layer thickness
p_p = permeability of polymer film
p_c = permeability of coating
p_l = permeability of laminate or other layer

If you check out the references relating to permeability it becomes apparent that there are a number of variations of the same basic equations used. This has been reviewed and a comparison made of the various equations (30) and the deficiencies of some highlighted. They also recommend the equations from one reference (31) as being useful.

2.1 Permeability Models

This ideal laminate theory equation for calculating the total permeability is often referred to as the electrical analogy model and the equation is in the same form as adding electrical resistors together to find the total resistance of a circuit. Figure 2.6 shows a schematic of the coating and permeability is varied with the three basic permeability models that are popularly used.

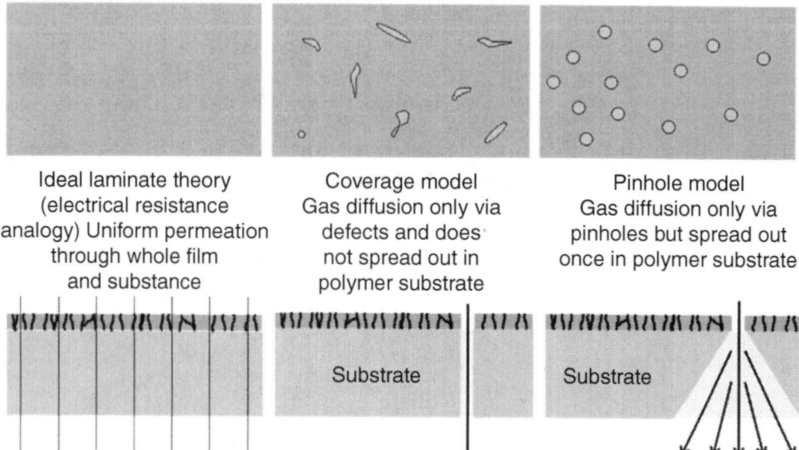

Figure 2.6 A schematic of the three most used model type for barrier permeability.

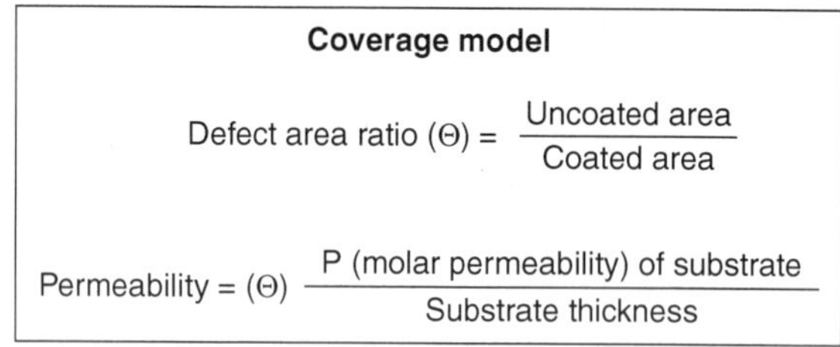

Figure 2.7 The modified permeability equation using the coverage model.

This value of total permeability of the multilayer will not be the same as that measured as it assumes there are no defects and the layers are homogeneous which is not the case in practice. This has resulted in alternative models for predicting the permeability of materials such as the coverage model or the pinhole model.

In the coverage model the area of the defects is taken into account to modify the coating permeability as shown in Fig. 2.7. The model assumes a unidirectional diffusion through the polymer from the exposed area of the polymer. This was deemed to be simplistic and the pinhole model is a variation where it is assumed that the diffusion will spread out in the polymer from the original defect size as shown in Fig. 2.8. Other assumptions made in the pinhole model are that there is no permeation through the thin film barrier material as the defects dominate the performance, all the defects are circular

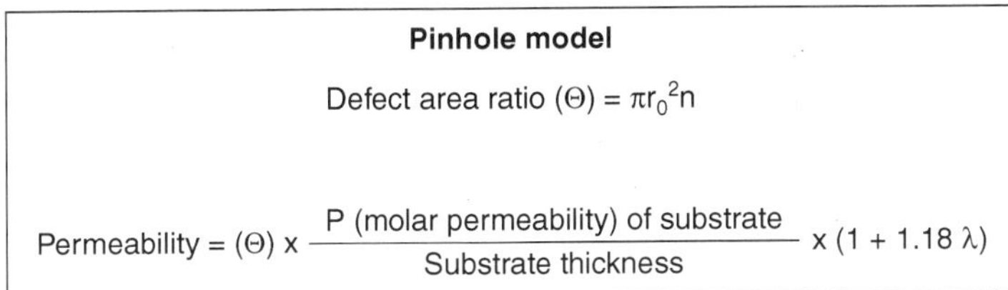

Figure 2.8 The modified permeability equation using the pinhole model.

and sufficiently far apart that they are additive and the pinholes are described by a mean radius and number per unit area (14).

These models have been examined and one unexpected result (17) that came out of the process highlights just why the pinholes are so damaging to the barrier performance of the vacuum deposited coatings. Calculations were done to find how much coverage of the substrate was required to have an effect on the permeability and it was found to be more than 95% before some improvement was seen. This assumes there was no permeation through the coating. If there were to be some permeation through the coating this figure becomes higher still. If we change this round we go out from perfect barrier performance of the coating with zero defects to zero barrier performance with less than 5% of the substrate area taken up with defects. This shows that the vacuum deposited coatings are enormously sensitive to defects and why pinholes are so damaging to the barrier performance.

For vacuum deposited coatings there is usually a critical thickness below which the barrier performance drops off rapidly. This thickness corresponds to the coating becoming continuous. Until the coating is continuous or in reality achieves coverage of greater than 95% it will have no barrier performance as per the coverage model. Above this critical thickness the diffusion is directly related to the coating thickness however there is also the component of the permeability that is independent of thickness and this relates to the major defects such as pinholes. These pinholes become the limiting factor as they prevent the coverage becoming 100%. The pinholes are more or less constant in size and have little or no barrier performance and are usually independent of coating thickness as many of them will have been produced after the vacuum deposition has been completed. Dust left on the polymer surface is coated and then the dust is rolled away after coating leaving behind the pinhole (32).

This is not the only source of pinholes, as will be seen later, but is the most common. The pinholes easily contribute the largest area of defects with pores, cracks, and grain boundary defects making up the minor contribution (14,16,33–48).

If we plot the pinhole diameter against the performance of the barrier coating as measured by the barrier improvement factor we can see the effect of pinholes on the coating performance. As soon as we have some pinholes, even of a small size the barrier performance drops very rapidly. This is shown schematically in Fig. 2.9 where the shaded area is to highlight this rapid fall off of barrier performance. This graph also shows that the pinhole model is much more sensitive to the problems caused by pinholes and so mimics reality compared to the coverage model which is less sensitive and affects the performance less.

This does highlight the problem of making improvements to the substrate quality so that the vacuum deposited coating can approach the theoretical performance. The very best ultra barrier coatings have the polymer substrate cleaned and then a polymer planarising layer added to cover up any remaining debris or surface damage to give the smoothest, flattest and cleanest surface possible. However unless the coating is completely pinhole free the few pinholes present will have a disproportionately huge effect in loss of barrier performance. As few processes can be guaranteed to be perfect every

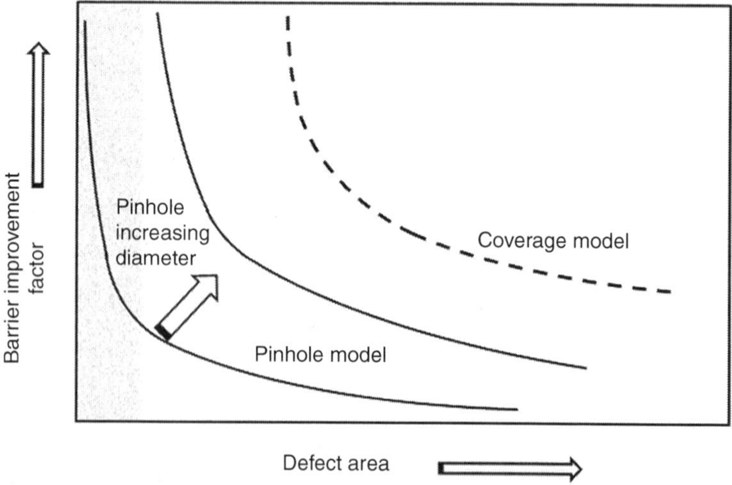

Figure 2.9 A schematic of the effect of defects on the barrier performance.

time it is common to use repeated smoothing layers and vacuum deposited barrier coatings to produce a composite barrier coating that can deliver a much higher barrier performance.

This does give some targets for improving the barrier performance of vacuum deposited coatings. Primarily, eliminate pinholes, this is followed by reducing or eliminating pores and finally densification of the coating and minimising defects such as grain boundaries. Finally once all these are under control the coating material can then be optimised as it is by no means certain that silica or alumina are the best materials for the vacuum deposited barrier coatings. It may be that they remain the most cost effective for the low cost applications but where the best possible performance is required it has been suggested that other materials may prove better (18,49).

Different deposition processes or substrate quality can give rise to a different coating structure and defect density. Thus it is not possible to give precise values of oxygen or water vapour permeability for specific vacuum deposited materials and thickness but only approximate values. It is therefore wise to regard most values as having a large tolerance associated with them.

Although we have these models to use they are more often than not used purely to explain why the practical values for permeability do not match the theoretical. The electrical analogy model is probably the most popular used to estimate the performance of a multilayer barrier structure.

In doing these calculations it is common for not all layers to be included with thin layers such as the adhesives being ignored. The argument used is that these layers are very thin and have little or no impact on the total performance and so they are ignored and are included in the error for the result. This may not be entirely true as the polymer adhesive may flow into the pinholes in the vacuum deposited coating and substitute adhesive for the air in the pinholes which would change the diffusion through the pinholes by several orders of magnitude. There have been comments in the past that laminating vacuum coated films can improve the barrier performance of the vacuum coated layer. It was thought that this was because the lamination prevented damage to the coating, which is true, but what may not have been understood is this filling of the pinholes and the effective change this could have on the diffusion through the pinholes.

In practice there can be many ways to achieve the same permeability goal and often not only is the total permeability calculated but also the cost for the package will be calculated and alternative structures compared (50).

2.2 Barrier Improvement Factor

Mentioned earlier and used in Fig. 2.5 is another measure that has been used to evaluate the benefits of using different deposition processes for the same material or to compare different materials. This measure is the Barrier Improvement Factor (BIF) (13,17). When using the ideal laminate theory the BIF is equal to the permeation rate of the uncoated polymer film divided by the permeation rate of the coated polymer film. When using the coverage model or the pinhole model the BIF has to be modified to take account of the inclusion of defects into the models.

By using the BIF it is possible to compare coatings of the same thickness and evaluate the deposition process or material to make the best or most cost effective choice to achieve the required barrier performance. Using this method of comparing the coatings it has been shown that sputtered alumina coatings are better oxygen and water vapour barrier coatings than electron beam deposited alumina coatings (51). This is as was expected as sputtered coatings are generally denser than evaporated coatings due to the higher energy of deposition.

There is another measure that has been used for comparison and that is the activation energy (52–56). This is the energy related to that required for a gap to be made between polymer chains to allow the absorbed molecule to pass through. This is found in the Arrhenius equation (57) for permeation;

$$P = P_0 \exp^{(-\Delta E/RT)}$$

where

ΔE = Apparent activation energy

R = Gas constant

T = Absolute Temperature

P_0 = Permeation constant unique to the system (related to the entropy of activation)

The activation energy has been evaluated for plain, single side coated and double side coated polymer films. All of the values were shown to be the same, within experimental error, which would suggest that there is no activation energy from either of the coatings and so they must be filled with defects leaving the activation energy for the polymer as the only one present (43). The same work shows that for water vapour there is some change in activation energy which would suggest that as the same number and size of defects would be present that there is some other interaction between the water vapour and the coatings although this is, as yet, not fully understood.

2.3 Tortuous Path Model

The simple view of gas diffusing through a material is that it passes in a straight line from one side to the other. Polymers are often mixtures of amorphous and crystalline regions. In pulling tension on the polymer to biaxially orient the polymer film more of the polymer chains are brought in close proximity to each other and nearer to parallel to the plane of the film. If heated these collections of closely packed polymer chains will form crystallites that are oriented in the direction of the tension. The crystalline regions are denser and it is easier for the gas to diffuse around the crystalline regions and pass through the amorphous regions as shown in Fig. 2.10. This diversion of the straight line path is the basis of the tortuous path model.

This same effect can be enhanced by adding fillers to the polymer matrix particularly if the filler has a high aspect ratio and is in the form of flakes rather than more spherical particles. These

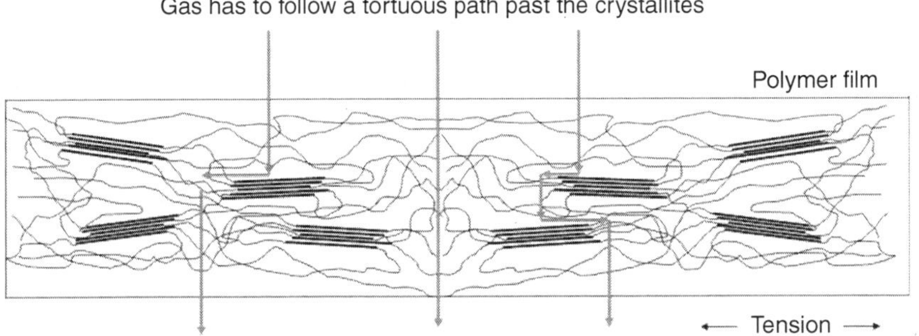

Figure 2.10 A schematic of how the crystallinity of a polymer film affects gas diffusion.

flake fillers can be used in the substrate, or added as a co-extruded layer or included in coatings added to the polymer web substrate as shown in Fig. 2.11.

If we look at what happens when we have a thin vacuum deposited coating on the polymer web substrate. We know that most coatings contain defects and so if we look at a model of the gas diffusion through a pinhole in the coating for two different thickness substrates we can see a schematic of the results in Fig. 2.12. Pinholes of the same diameter show a restriction in the lateral diffusion with the thinner substrate. The pinholes of smaller diameter also show a reduced lateral diffusion.

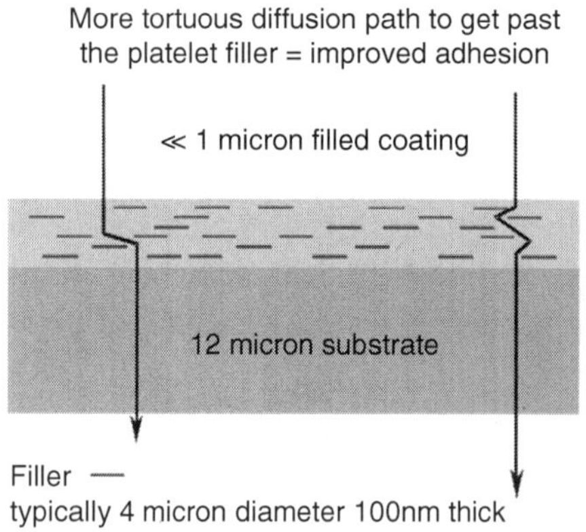

Figure 2.11 Extending the tortuous path model can be done using flake fillers.

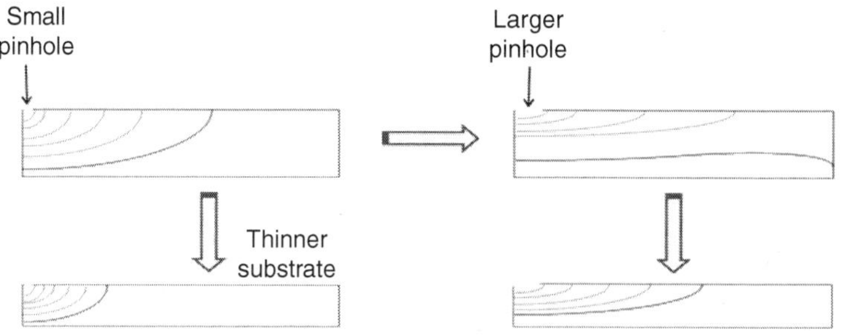

Simulation of diffusion of gas through pinholes – red lines are for equivalent concentration

Larger pinholes – more gas diffuses through and to a wider position in the substrate
Larger pinholes – spread of diffusion means pinholes have to be further apart to be independent
Thinner substrate – limits the spread of diffusion from the pinhole

Figure 2.12 The effect on substrate thickness and pinhole size on diffusion.

This information explains why double side coating can be so beneficial in improving barrier performance. If we look at Fig. 2.13 we can see that if both sides are coated then a defect in one coating is unlikely to be directly opposite a defect in the coating on the other side. If the defect density is low and the defect spacing is much larger than the substrate thickness then the lateral diffusion to a defect on the opposite side is much lower than the diffusion would be straight through the substrate. If the defect density is high in both coatings the lateral diffusion distance is much lower and so the benefits are not as noticeable.

It is possible to exaggerate this tortuous path model by depositing very thin polymer layers to separate alternating layers of high vacuum deposited barrier materials. This is shown schematically in Fig. 2.14. In this structure the initial polymer coating smoothes out the polymer substrate surface and aims to minimise the number of pinhole defects and filler induced defects. Once the first transparent ceramic or glass-like barrier coating is deposited another polymer layer is deposited over the top and this serves two purposes. Firstly, any defects are filled with polymer and so the diffusion rate through the pinholes then reduces from the diffusion rate in air to the diffusion rate in the coating polymer. The second purpose is that the polymer separates the first transparent inorganic coating from the second inorganic coating. This polymer coating can be very thin and so the lateral diffusion between defects can be significantly reduced. This

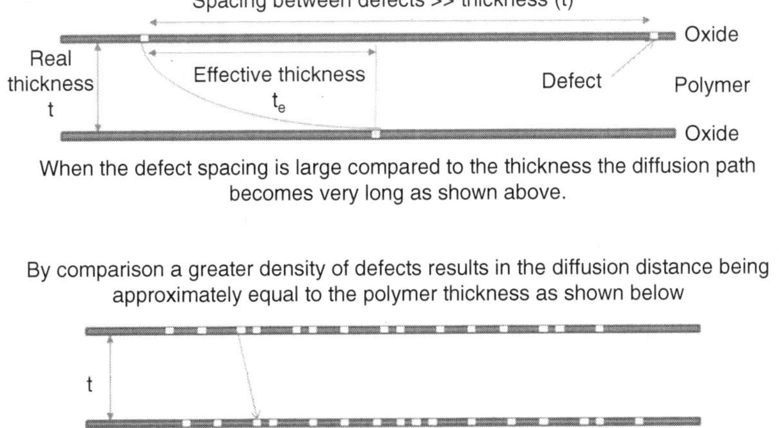

Figure 2.13 The effect of double side coating and defect density on permeability.

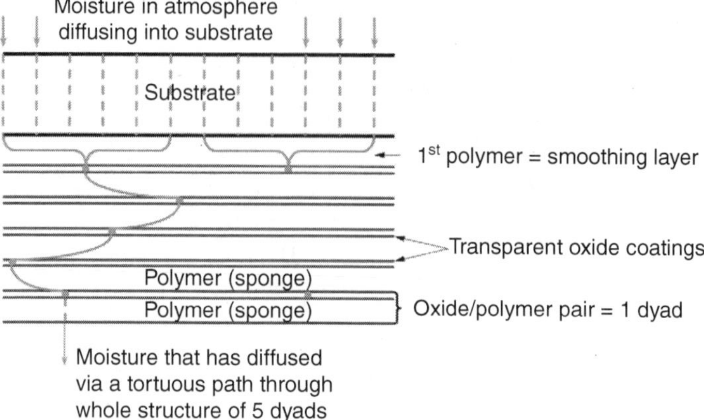

Figure 2.14 A schematic of the effect of alternating polymer and high barrier transparent vacuum deposited coatings.

process of alternating organic and inorganic layers can be repeated a number of times to ensure the total permeability is reduced to the desired level. A pair of coatings comprising an organic and an inorganic coating together are sometimes referred to as a dyad. Hence a 5 dyad barrier structure comprises 10 separate coating layers. There will still be diffusion through the whole structure but it will be very low which means that the time to test these structures can become very long. Early references claiming ultra barrier performance were found to be untrue as the testing time had not been done for long enough and the structures had not reached an equilibrium state. Once the same structures were tested for long enough the permeation rate increased once the structure had become saturated. Think of the structure as a sponge where of you pour water on the sponge, the sponge will keep absorbing the water until it becomes saturated at which point water will start pouring out of the other side.

2.4 Terminology Summary

Permeation – the total process of absorption + diffusion + desorption.

Sorption – also known as absorption and also known as scalping – is the process of gas transferring from the free gas surrounding the packaging material into the barrier material.

Desorption – this is the opposite process to absorption – the gas that has passed through the coated polymer barrier material is released into the free gas adjacent to the barrier material surface.

Diffusion – the process of the absorbed gas moving through random motion through the barrier material. Diffusion is in all directions but can be given a direction if there is a concentration or pressure gradient with the gas moving from the high concentration or high pressure region to the low concentration or low pressure region.

Migration – this is the process of volatile material being lost from the barrier packaging material into the product being packaged. The possibility is that this migrating material can cause taint or toxicity problems in some foodstuffs.

Activation energy – an indication of the resistance of the material to permeation.

BIF – Barrier improvement factor – this is a measure of how much better the polymer film plus coating is against the original substrate polymer film.

Tortuous path – where the easy diffusion path is not a straight line but deviates around regions of high barrier material.

Dyad – a term that refers to a pair of layers comprising adjacent organic and an inorganic coatings.

References

1. Konkol L. 'Chapter 2.2 – Food contact considerations for virgin PET' In Thesis *'Contaminant levels in recycled PET'* Nov 2004 pp 5–9 Environment & Biotechnology Centre, Swinburne Univ. of Tech. Victoria, Australia.
2. Paik J.S. 'Comparison of sorption in orange flavour components by packaging films using headspace techniques' *J. Agric. Food Chem.* 40, 1992 p 1822 1825.
3. Tavss E.A. et al. 'Analysis of flavor absorption into plastic packaging materials using multiple headspace extraction gas chromatography' *Journal of Chromatography A* Volume 438, 1988, pp 281–289.

4. Nielsen T.J. and Olafsson G.E. 'Sorption of β-carotene from solutions of a food colorant powder into low-density polyethylene and its effect on the adhesion between layers in laminated packaging material' *Food Chemistry* Vol. 54, Issue 3, 1995, pp 255–260.
5. Gavara R. et al. 'Study of aroma scalping through thermosealable polymers used in food packaging by inverse gas chromatography'. *Food Additives & Contaminants* 14, 1997 pp 609–616.
6. Nielsen T.J. et al. 'Comparative Absorption of Low Molecular Aroma Compounds into Commonly Used Food Packaging Polymer Films' *Journal of Food Science*, 57, 1992, pp 490–492.
7. Nielsen T.J. and Jägerstad I.M. 'Flavour scalping by food packaging' *Trends in Food Science & Technology* Vol 5, Issue 11, Nov. 1994, pp 353–356.
8. Tice P.A. and McGuinness J.D. 'Migration from food contact plastics. Part I. Establishment and aims of the PIRA project' *Food Additives & Contaminants: Part A Chemistry, Analysis, Control, Exposure & Risk Assessment*, Vol. 4, Issue 3, 1987, pp 267–276.
9. Boone J. 'Deficiencies of polypropylene in its use as a food-packaging material - a review' *Packaging Tech. & Sci.* Vol 6, Issue 5, 1993, pp 277–281.
10. Crosby N.T. '*Food packaging materials – Aspects of analysis and migration of contaminants*' Applied Science Publishers 1981, pp 106–149.
11. Sing K.S.W. et al. 'Reporting physisorption data for gas/solid systems - with special reference to the determination of surface area and porosity' *Pure & Appl. Chem.*, Vol. 57, No. 4, 1985, pp 603–619.
12. Suloff E.C. 'Chapter 4 – Permeability, diffusivity and solubility of gas and solvent through polymers' in Thesis '*Sorption behaviour if an aliphatic series of aldehydes in the presence of poly(ethylene terephthalate) blends containing aldehyde scavenging agents*'. Virginia Polytechnic Institute and State University Nov 2002 pp 29–99.
13. Chatham H. 'Review – oxygen diffusion barrier properties of transparent oxide coatings on polymeric substrates' *Surface and Coatings Technology* 78, 1996, pp 1–9.
14. Prins W. and Herman J.J. 'Theory of permeation through metal coated polymer films' *J. Phys. Chem.* 63 (5), 1959, pp. 716–719.
15. Gudimov M.M. et al. 'Effect of metallization of polymer films on their gas permeability' Translated from *Fiziko-Khimicheskaya Mekhanika Materialov*, Vol. 7, No. 2, March–April, 1971, pp. 59–63.
16. Beu T.A. and Mercea P.V. 'Gas transport through metallized polymer membranes' *Mat. Chem. Phys.* 26, 1990, pp 309–322.
17. Decker W. and Henry B.M. 'Basic Principles of Thin Film Barrier Coatings' Soc. Vacuum Coaters 45th Ann. Tech. Conf. Proc. 2002, pp 492–502.
18. Henry B.M. 'Permeation Studies of Transparent Barrier Coatings' Soc. Vacuum Coaters 46th Ann. Tech. Conf. Proc. 2003, pp 600–605.

19. Moosheimer U. and Langowski H-C. 'Permeation of oxygen and moisture through vacuum web coated films' Soc. Vacuum Coaters 42nd Ann. Tech. Conf. Proc. 1999, pp 408–414.
20. Moosheimer U. 'Permeation of Oxygen, Water Vapor and Aroma Substances Through Vacuum Web Coated Films' Soc. Vacuum Coaters 43rd Ann. Tech. Conf. Proc. 2000, pp 385–390.
21. Pauly S. 'Permeability and diffusion data' in *'Polymer Handbook'* 4th Edn. Eds. Brandrup J. Immergut E.H. Grulke E.A. John Wiley & Sons Inc.: NY, 1999, pp 543–569.
22. Beu T.A. and Mercea P-V. 'Gas transport through metallized polymer membranes' *Materials Chemistry and Physics* 26, 1990, pp 309–322.
23. Schrenk W.J. and Alfrey Jr. T. 'Some physical properties of multi-layered films' *Polymer Engineering and Science* Vol. 9, Issue 6, 1969, pp 393–399.
24. Brody A.L. et al. 'Innovative food packaging solutions' *J. Food Sci & Tech* Vol. 73, No. 8, 2008, pp R107–R116.
25. DeLassus P.T. and Strondberg G. 'Flavor and aroma permeability in plastics' in 'Food packaging technology' Ed. Henyon D. American Soc. For Testing Matls. ASTM STP 1113, 1991 pp 64–73.
26. Van Willige R.W.G. *'Effects of flavour absorption on foods and their packaging materials'* Thesis Wageningen University, The Netherlands, 2002.
27. Utto W. *'Mathematical modelling of active packaging systems for horticultural products'* Thesis Massey University, New Zealand, 2008.
28. Delassus P.T. 'Barrier polymers' in *'The Wiley Encyclopedia of packaging technology'* 2nd Edn. Eds Brody A.L. and Marsh K.S. John Wiley & Sons, Inc., New York, 1997, pp 71–77.
29. 'Non-Oxygen Conventional Gases Permeability Analysis' Customer factsheets, Labthink Instruments Co.,LTD, China www.labthink.cn/service/show987.html.
30. Cooksey K. et al. 'Predicting permeability and transmission rate for multilayer materials' Food Technology. 53 (9), 1999, pp 60–63.
31. Robertson G.L. *'Food Packaging Principles and Practice'* Publisher Marcel Dekker, Inc., New York 1993, p100.
32. E.H.H. Jamison & A.H. Windle 'Structure and oxygen barrier properties of metallized polymer films' *J.Matls.Sci.* Vol. 18, pp 64–80, 1983.
33. Weiss J. et al. 'Al-Metallization of Polyester and Polypropylene Films: Properties and TEM Microstructure investigations' *Thin Solid Films,* 174 (1989) pp 155–158.
34. Weiss J. 'Parameters that influence the barrier properties of metallized polyester and polypropylene films' *Thin Solid Films,* 204 (1991) pp 203–216.
35. Rossi G. and Nulman M. 'Effect of local flaws in polymeric permeation reducing barriers' *J. Appl. Phys.* 74, 1993, pp 5471–5483.

36. Henry B.M. et al. 'Microstructural and gas barrier properties of transparent Aluminium Oxide and Indium Tin Oxide Films' 43rd Ann. Tech. Conf. Proc. Society of Vacuum Coaters 2000, pp 373–378.
37. Henry B.M. et al. 'Characterization of oxide gas barrier films' 42nd Ann. Tech. Conf. Proc. Society Vacuum Coaters, 1999 pp 403–407.
38. Erlat A.G. et al. 'Characterisation of Aluminium Oxynitride gas barrier films. *Thin Solid Films,* 388, 78, 2001, pp 78–86.
39. Henry B.M. et al. 'The permeation of water vapour through gas barrier films' 44th Ann. Tech. Conf. Proc. Society Vacuum Coaters, 2001 pp 469–475.
40. Hanika M. et al. 'Simulation and verification of defect-dominated permeation mechanisms in multiple structures of inorganic and polymeric barrier layers' 46th Ann. Tech. Conf. Proc. Society Vacuum Coaters, 2003 pp 592–599.
41. Roberts A. P. et al. 'Gas permeation in silicon-oxide/polymer (SiOx/PET) barrier films: role of the oxide lattice, nano-defects and macro-defects' *J. of Membrane Science,* Vol 208, Issues 1-2, 1 Oct 2002, pp 75–88.
42. Da Silva-Sobrinho A.S. et al. 'A study of defects in ultra-thin transparent coatings on polymers'. *Surface & Coatings Technology,* 119, 1999, pp 1204–1210.
43. Da Silva-Sobrinho A.S. et al. 'Defect-permeation correlation for ultrathin transparent barrier coatings on polymers' *Journal of Vacuum Science & Technology A,* 18(1), 2000, pp 149–157.
44. Yanaka A.I. et al. 'How cracks in SiOx-coated polyester films affect gas permeation' *Thin Solid Films,* 397(1-2), 2001, pp 176–185.
45. Garcia-Ayuso G. et al. 'Relationship between the microstructure and the water permeability of transparent gas barrier coatings' *Surface and Coatings Technology,* Vol. 100-101, No 1-3 1998, pp 459–462.
46. Hanika A.I. et al. 'Inorganic layers on polymeric films - influence of defects and morphology on barrier properties. *Chemical Engineering & Technology,* 26(5), 2003, pp 605–614.
47. Czeremuszkin G. et al. 'A simple model of oxygen permeation through defects in transparent coatings' 42nd Ann. Tech. Conf. Proc. Society Vacuum Coaters, 1999, pp 176–180.
48. Alueller K. and Weisser I.I. 'Numerical simulation of permeation through vacuum-coated laminate films' *Packaging Technology and Science,* 15(l), 2002, pp 29–36.
49. Henry B.M. et al. 'Characterization of transparent aluminium oxide and indium tin oxide layers on polymer substrates' *Thin Solid Films,* Volume 382, Issues 1–2, 1 Feb 2001, pp 194–201.
50. Butler T.I. and Morris B.A. 'PE based multilayer film structures' in Chapter 15 of *'Multilayer flexible packaging'* Ed. Wagner Jr. J.R. Publisher Elsevier 2009, pp 209–212.

51. Henry B.M. et al. 'Permeation barrier studies of multilayer films' 48th Ann. Tech. Conf. Proc. Society Vacuum Coaters, 2005, pp 644–648.
52. Klute C. H. and Franklin P. J. 'The permeation of water vapor through polyethylene' *J. of Polymer Science*, Vol. 32 Issue 124, 1958, pp 161–176.
53. Erlat A.G. et al. 'SiOx gas barrier coatings on polymer substrates: morphology and gas transport considerations' *J. Phys. Chem. B*, 103 (29), 1999, pp 6047–6055.
54. Howells D.G. et al. 'High quality plasma enhanced chemical vapour deposited silicon oxide gas barrier coatings on polyester films' *Thin solid films* Vol. 516, No10, 2008, pp. 3081–3088.
55. Henry B.M. et al. 'A microstructural study of transparent metal oxide gas barrier films' *Thin Solid Films*, Vol. 355–356, 1 Nov. 1999, pp 500–505.
56. Tropsha Y.G. and Harvey N.G. 'Activated rate theory treatment of oxygen and water transport through silicon oxide/poly(ethylene terephthalate) composite barrier structures' *J. Phys. Chem. B*, 101 (13), 1997, pp 2259–2266.
57. Fahlteich J. et al. 'Mechanical and barrier properties of thin oxide films on flexible polymer substrates' 51st Ann. Tech. Conf. Proc. Society Vacuum Coaters, 2008, pp 808–813.

3

Measurements

There are two groups of measurements that we shall go through, those that can be done off-line and those that can be within the roll-to-roll vacuum deposition process. As the focus of this book is the barrier performance of our coatings we shall start with a number of options for testing the barrier performance of the coated films. Coupled to these barrier tests there are also a number of tests that are done that simulate some of the handling or packaging mechanisms. Following these simulated processes the barrier performance can be re-tested and the ability of the coated barrier materials to withstand the different mechanisms evaluated. Some of these tests are as simple as bending and require little equipment and technical skills whereas others can require very sophisticated equipment and technical skills.

As we know that the surface quality and cleanliness of the substrate has a large impact on the barrier performance we shall also include tests that are used to evaluate these properties. Some of these tests can be done before the coating process whereas others, such as counting pinholes, can only be done after the coating process.

3.1 Permeation Measurements

There are different methods of measuring the permeation through barrier coated polymer films. All of the methods require a sample of the material to be tested to be mounted to divide two volumes and the quality of this seal becomes one of the limiting factors of the test method. The permeability measured can be the combination of gas permeating through the sample plus any gas leaking through the seal. The lower the permeability of the sample the more sensitive the measurement is to seal leaks.

One popular method is to have two chambers separated by the barrier film to be tested. Both chambers are at the same pressure but the gas composition of each chamber is different. In one chamber the gas for which the permeability value is sought is passed through and in the other dry nitrogen is passed through. The concentration difference between the two chambers will drive the gases to pass through the barrier material. Nitrogen will pass through in one direction and the sample gas will pass through into the dry nitrogen chamber. The exhaust from the dry nitrogen chamber passes through a sensor which is sensitive to the test gas. This is shown schematically in Fig. 3.1. A coulometric sensor (1) based on cadmium is used for oxygen where the sensor works on the reverse fuel cell principle & allows the oxygen content to be converted into an electrical signal. There are also zirconia based detectors also for use with oxygen (2). Once a new sample is mounted between the two cells the system is flushed out to make sure any contamination is minimized and then the gases are introduced. The test is completed when the concentration of the test gas in the nitrogen gas reaches a constant level and the process is in a steady state. The gases are controlled both in temperature and humidity. The dry nitrogen is always at 0% relative humidity (RH) and at a controlled temperature. The sample gas will be at the same temperature but may be at a different relative humidity level. This same methodology can be used to measure the water vapour permeability too by using nitrogen on both sides of the cell with the detection nitrogen being dry at 0% RH and the sample gas side being at 100% RH. This measurement is designed to meet ASTM F1249 (2). The detection system for water uses an infrared sensor. There are alternative detection systems (3) such as the gravimetric sensor for water vapour. There is not a single universal sensor and so these machines are built specifically for each sample gas which is

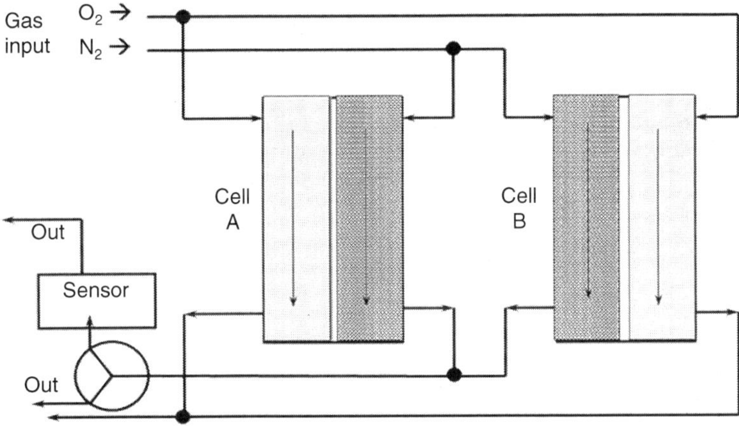

Figure 3.1 A schematic of the 'Oxtran®' type of gas permeability measurement system (5).

why it is sometimes the case that estimates of the permeability for some gases are used by multiplying the known gas permeability by some factor. This can result in a value that has a very high error as the multiplication factor will change with temperature and polymer substrate type but often only a single multiplication factor is given as a universal factor. Thus this way of estimating the permeability should be used with caution. This type of permeability measurement system has been developed and can measure the water vapour transmission rate down to 0.0005 gm/m^2/day and the oxygen transmission rate down to 0.0005 cc/m^2/day. These minimum values can be verified and traced back to standards at the National Institute of Standards and Technology (NIST), USA. The most widely quoted systems of this type are the Ox-tran® and Permatran® or Aquatran® from Mocon (4). It will be noted that the minimum values that these systems can measure are still at least a couple of orders of magnitude short of the sensitivity necessary to be able to measure the permeability of the ultra barrier materials being developed for the OLED and photovoltaic applications.

A similar system that uses a pressure difference between the two sides of the barrier film can also be used. In this case the pressure rise with time can be measured and as shown in Fig. 3.2 the permeability and diffusion can be determined and using these two values the solubility can be calculated.

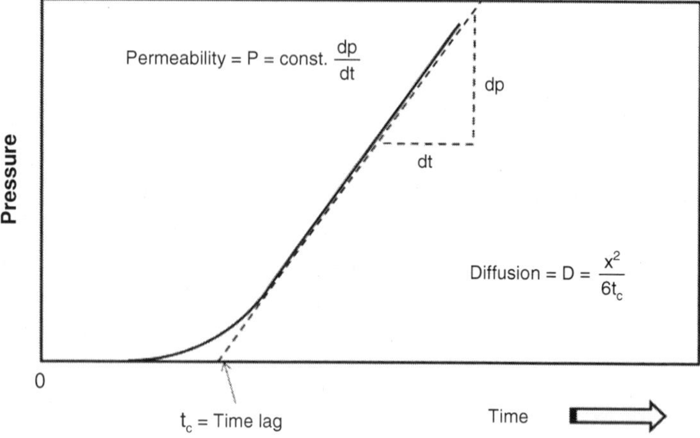

Figure 3.2 A plot of pressure change with time provides a measure of permeation and diffusion.

A variation of this technique is to use a gas chromatograph instead of the simple pressure gauge. This system does allow for gas mixtures to be measured.

Newer systems have tried to optimise existing systems by reducing the ingress of contaminants by improving the joints and seals to minimise the leaks. This reduction in gas ingress thus improves the minimum quantity of transmitted gas or vapour that can be reproducibly detected. Bearing in mind that it is common to have a test method that measures at least one decade better than the expected value of item it would suggest that this technique needs to be improved by three orders of magnitude before it will be suitable for use on ultra barrier materials.

Figure 3.3 shows the use of a cell within an ultra high vacuum system (6–9). The sealed cell has a drop of water as the moisture source & anything leaking through the barrier coating is detected using a mass spectrometer. The drop of water provides the cell with a constant partial pressure of water vapour. Isotope $^{20}H_2O$ has been used to improve the sensitivity of this method for measuring water vapour permeability. This isotope enables the water permeating to be differentiated from any water vapour outgassing from the system. The mass spectrometer measures the partial pressure of water vapour which is proportional to the water vapour transmission rate. If the source of gas is sealed off the decay of the detected water vapour will give a measure of the outgassing rate of the polymer substrate.

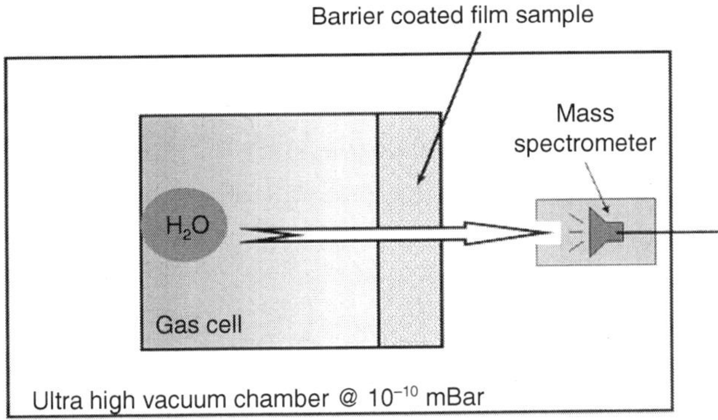

Figure 3.3 A schematic of a vacuum based permeability measurement system.

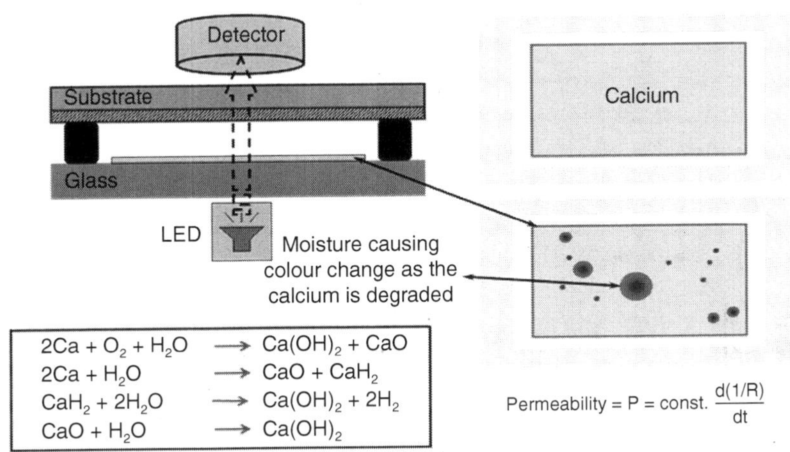

Figure 3.4 A schematic of the calcium test for ultra barrier coating materials.

With a suitable source of gas this can be used for other gases. This has the potential for delivering a reproducible answer and is expected to be capable of measuring down to the levels required by the display industry. The sources of error are from seal leaks or outgassing of the surfaces in the vacuum. As the mass spectrometer can detect all gases the single system has the potential to measure any gas or mixtures of gases.

The most reported method of measuring the permeability levels required by the display or photovoltaic industries is the Calcium test. Figure 3.4 shows a schematic of the Calcium corrosion test (10–16). Here the calcium is coated on the coated barrier film & sealed with the sample of the barrier film separated from a glass plate using a gasket. A light is directed through the calcium, which is opaque, to a detector. As the calcium corrodes, as per the reactions shown in Fig. 3.4, it becomes transparent & any changes follow the Beer-Lambert Law and may be measured by the detector. Alternatively the conductivity of the calcium can be detected directly by a change in resistivity and from this change in resistivity with time the permeability can be determined from the slope as give in the equation in Fig. 3.4.

This test is only really applicable to moisture barrier performance. This test does appear to be very operator sensitive requiring a high degree of expertise to obtain any degree of reproducibility. Often the calcium areas are arranged in arrays of 2 × 2, 3 × 3 or 4 × 4 to give some measure of sample and calcium coating uniformity. As this test is so sensitive to the skill of the technician running the test it is prudent for multiple samples to be prepared to ensure there is a reasonable confidence the values are accurate. There is not an off-the-shelf test system that can be bought. The calcium is so sensitive that it has to be deposited onto the sample and used immediately as it cannot be stored. There is a high degree of skill required to be able to deposit and assemble the calcium layer into a test cell in a clean dry atmosphere so that the test material is not contaminated before the test starts. Hence this is regarded very much as a research tool rather than a more widely applicable industry test that can be used by every operator or technician. With this high degree of preparation and skill required there are now a few laboratories offering this test service.

The degradation of the calcium starts from defects and so as the series of degradation spots appear it becomes a way of estimating the number of defects per unit area which in turn can be used as a metric to measure any coating or process improvement.

One group in Japan have opted to develop an alternative to the calcium corrosion test to make a much easier test. In this case cobalt chloride is sealed in a bag made of the barrier material and the weight of the bag monitored. Any water ingress will be absorbed by the cobalt chloride and the weight increase in the bag over time is used to determine the permeability. Although this was a much easier test to prepare and required a much lower level of operator skill it had the limitation that it could take several weeks (months for good samples) to produce sufficient data points to get a graph to provide the permeability value. To make the test more sensitive and quicker to deliver a result the colour change as the cobalt chloride reacts with the water was monitored. A cobalt chloride solution was soaked into a paper that could be laminated onto the barrier film and the colour, as it changed from blue to red, was monitored in detail (17). The rate of colour change was then used to determine the permeation rate. The chemical reaction that gives rise to the colour change is as follows;

$$\underset{\text{(blue)}}{CoCl_2} + nH_2O \rightarrow \underset{\text{(pale blue)}}{CoCl_2{\cdot}nH_2O} + (m\text{-}n)\,H_2O \rightarrow \underset{\text{(red)}}{CoCl_2{\cdot}mH_2O}$$

This variation on the bag test did improve the response time by a factor of 18x for barrier materials that they classed as ultra barrier materials.

The early or original use of calcium was in devices where it was a material that needed to be protected by using barrier materials. Then, because of its sensitivity, it was realised that it could be used as the test marker material. Similarly it is also possible to use lithium in the same way (18). With lithium the weight change due to oxidation can be plotted to provide the permeation information. The problems of working with lithium offer no advantage over working with calcium and so although possible this has not been as popular as the calcium test.

The organic light emitting display (OLED) industry have indicated that their target for the lifetime of displays is a minimum of 10,000 hours. The chemicals used are sensitive to moisture and it has been calculated that the rate of moisture ingress requires a water vapour transmission rate (WVTR) value of 1×10^{-6} g/m^2/day. It is this value that is now used as the unofficial definition with an ultra barrier material being one that can achieve this performance. With some OLEDs an aluminium electrode is used and it has been evaluated, from accelerated testing, that a WVTR value of 1×10^{-4} g/m^2/day would mean that a 200 nm thick aluminium coating would have had 20% of the thickness oxidised in 3 years and 67% in 10 years. Improving the barrier to a WVTR value of 1×10^{-5} g/m^2/day would reduce the aluminium corrosion to approximately 5% in 10 years.

One difficulty in measuring this is that there is no standard that can be used to calibrate tests and confirm quoted values. The current standard is more than a couple of orders of magnitude above the permeation levels that have to be measured. The calcium test has the ability to measure down to the desired level but is not

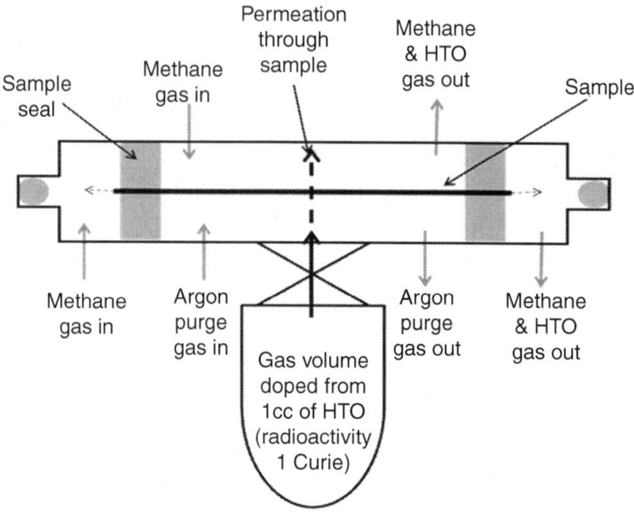

Figure 3.5 A schematic of the tritium doped permeability measurement system.

expected to be suitable to be used as the new standard. The use of a system with two separate volumes separated by the ultra barrier sample can be used but with tritium doped water in order to increase the sensitivity of detection down to 1×10^{-8} g/m^2/day. As with the other tests this does require that the sealing of the sample is not a source of contamination. In Fig. 3.5 we can see that there is the possibility of permeation through the sample and across the seal. This can be measured separately to the main permeation taking place normally through the bulk of the sample. The detection of the tritium doped water continues until a steady state value of the permeation is produced. This only occurs once the barrier coated material is saturated. Until saturation is reached the permeation will appear lower than the true value. As the water needs to be tritium doped it requires that laboratories wanting to work with this material obtain a special licence to work with radioactive materials and have implemented the necessary safety measures. This means that this technique can measure to at least one order of magnitude lower than necessary for the ultra barrier materials, which is better than for the calcium test, but is not easily available to most research and development groups (19–22). However it does provide a route for the standards bodies to produce a new standard with which they can calibrate the calcium test. This process of evaluation and defining new permeation standards is in progress in a number of countries. Until this process of upgrading the standards is completed there will be an uncertainty to some of the permeation values quoted for some of the ultra barrier materials

For all of these techniques it is important that the barrier samples are handled carefully. It has been shown (23,24) that simply light brushing of the surface can produce pinholes as any protruding debris is moved. As some of the vacuum deposited coatings may be ceramic and quite brittle it is also possible that cutting the samples will produce cracking on the coating and any deflection of the coated sample, including by clamping of the substrate in the test machine, can also lead to cracking. Any of these induced defects will degrade the permeation values for the samples.

3.2 Durability Testing

Coupled to the measurement of permeation through the barrier samples are a number of other tests that are aimed at differentiating the samples that are robust from those that are not.

An example of this would be the Gelbo Flex Durability Tester which is designed to meet ASTM test method F392 that tests the durability of coated polymer films to flexing. The sample is held between two rotating shafts so that each end of the sample is rotated in the opposite direction so that the sample is twisted as shown schematically at the top of Fig. 3.6. The direction of rotation is then reversed and this process is repeated a number of times. This can be a severe test and is more typical of the food packaging industry than it is of the display or photovoltaic industries. Where the sample is flexed or rubs against itself the coating can be damaged by producing pinholes if any protruding particles are moved. These moved particles may also produce scratches as the particles are rubbed across the surface. The test conditions aim at producing more than 5 but less than 50 pinholes and no other damage. It is not aimed at producing abrasion damage.

Once the sample has been through this test the number of pinholes can be counted and compared to samples before the test. Also the samples can be tested for permeation although if too many pinholes are produced this can reduce the barrier performance to the point where it is really only testing the polymer substrate.

Similarly the coated polymer films can be bent, with the coating side outwards, around solid mandrels of reducing diameters and examined for cracking. As the mandrel diameter is reduced the stress

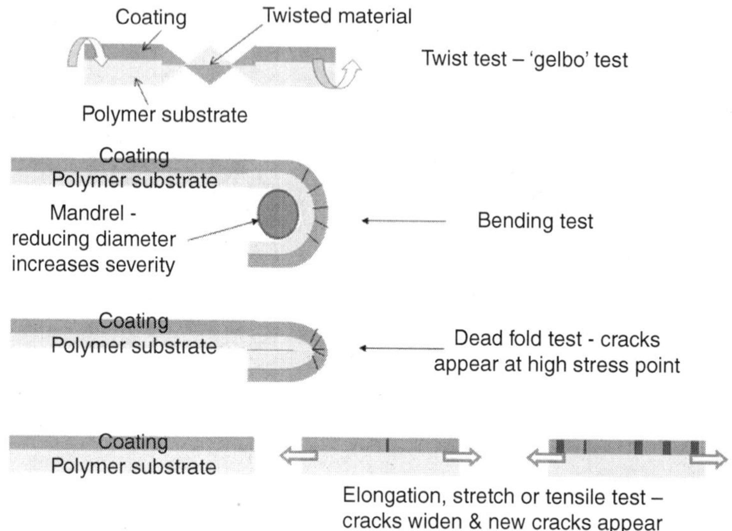

Figure 3.6 Schematics of various tests that can indicate how robust the coating is.

on the coating is increased. The coated film is bent round the mandrel and examined for cracking. The higher the adhesion between the coating and the polymer substrate the smaller the mandrel diameter it will be possible to bend the coated barrier polymer around without any sign of cracking. Where the substrates are stiff enough then a two point bending test can be used in a similar manner (25). If the adhesion of the coating to the substrate is poor the cracking will begin at much lower stress levels around larger mandrel diameters.

The ultimate bending test is the 'dead fold' test where there is no mandrel and the coated sample is simply folded and creased. This is a practical test for food packaging barrier materials where the packaging material is used to make up rectangular containers such as the vacuum packed packs of coffee. It is known that this type of handling of the barrier materials can cause a degradation in the barrier performance and so the specification of the barrier performance is often raised to allow for the performance to fall back during downstream processing (26,27).

The vacuum deposited barrier coating can be under stress from the deposition process. The type and degree of the stress will depend on the material deposited and the deposition process. Simple evaporated coatings are less dense and usually under less stress than coatings deposited by magnetron sputtering or other plasma enhanced deposition processes. A simple method of evaluating the coating stress is to cut a small rectangle of coated polymer film and to let it stand on edge so that it is allowed to bend as it wishes. Then using the formula originally developed by Stoney for metallic coatings deposited by electrolysis (28) the stress in the coating can be calculated. It is necessary when using biaxially oriented polymer films to cut the rectangles along a neutral axis to prevent substrate stresses altering the bend resulting from the coating.

Figure 3.6 also shows at the bottom a tensile test where the sample is pulled and monitored for cracks. The durability of the barrier coating depends on the adhesion between the coating and polymer substrate which affects delamination, the stress transfer between the substrate and coating and the cohesion in the coating which determines cracking (29–35).

The higher the adhesion the more of the load can be shared between the substrate and coating. The cracking of the coating acts as a stress relief mechanism. Using the micro-tensile testing stage it is possible to produce saturation of stress cracking from which the interfacial adhesion can be calculated. The measurement of the cracks can be made with more precision by using atomic force microscopy (AFM) where the stylus is dragged in a direction normal to the cracks. Figure 3.7 shows how increasing the tensile loading of the coated polymer films results in cracking of the coating.

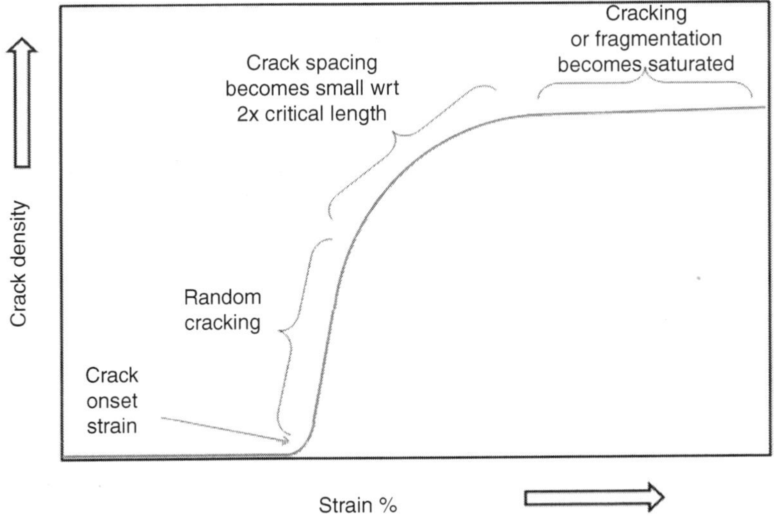

Figure 3.7 A plot of strain vs. crack density showing the three stages of cracking.

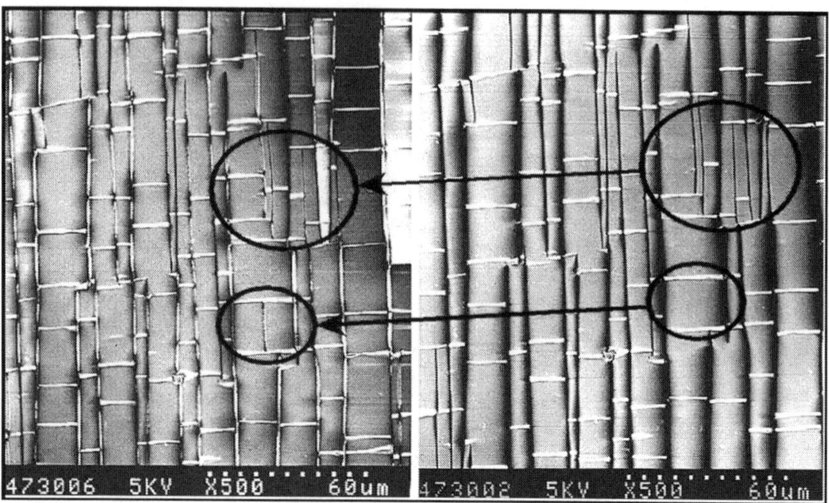

Figure 3.8 Scanning electron micrographs showing cracking typical after tensile testing.

Figure 3.8 gives an example of a coated polymer film that has been strained and is getting close to reaching crack density saturation. The highlighted rings are to help see where additional cracks have appeared with increasing load. The micrographs show the cracks in one direction and the coating buckling normal to the crack orientation due to the compressive stress the film experiences at 90 degrees to the tensile stress.

The other piece of information that this tensile testing technique gives is that of the onset of cracking (35–39). There is interest in how much stretch can be withstood by the coated barrier material before cracks appear as this is an indication of how robust the coating will be to handling and deformation during packaging processing. In this case the elongation at which point the first cracks appear is of most interest. The samples are stretched and the permeation measured and the onset of cracking corresponds to the change in slope of the permeation performance where there begins to be a huge loss in barrier performance. This work has primarily been done on the ceramic or glass-like coated barrier materials where the coating is brittle in comparison to the polymer substrate. The work has also showed some correlation in the substrate quality and the onset of cracking. The smoother and more perfect the polymer surface the higher the strain before the onset of cracking. Any defects in the

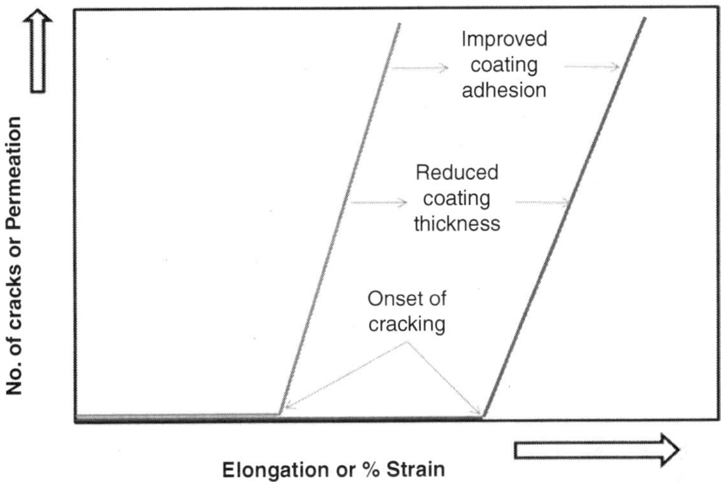

Figure 3.9 A schematic of strain vs. permeation showing the onset of cracking.

coating or substrate can become crack initiators and so the lower the quantity of filler protruding through the polymer surface or debris on the surface that leads to pinholes the higher the expected onset of cracking. The adhesion to the polymer surface is also critical and for the same surface quality the higher the adhesion the higher the strain before cracking starts. The onset of cracking is also improved as the coating thickness is reduced and the brittle coating becomes more flexible as shown schematically in Fig. 3.9. Once a substrate quality and surface preparation process, such as cleaning or corona or plasma treatment, has been established the onset of cracking can become one measure of reproducibility.

3.3 Adhesion

One of the most universally used tests for testing adhesion is the Scotch® or Sellotape® tape test. The test is named after the brand of adhesive tape used to perform the test and is shown schematically in Fig. 3.10. The method for the test is set out in ASTM D3359 - 09 'Standard test methods for measuring adhesion by tape test'.

Although this is a quick and easy test to perform it is really only a test of very poor adhesion and tells very little about the coating adhesion. The test requires the adhesive tape to be pressed onto the coated

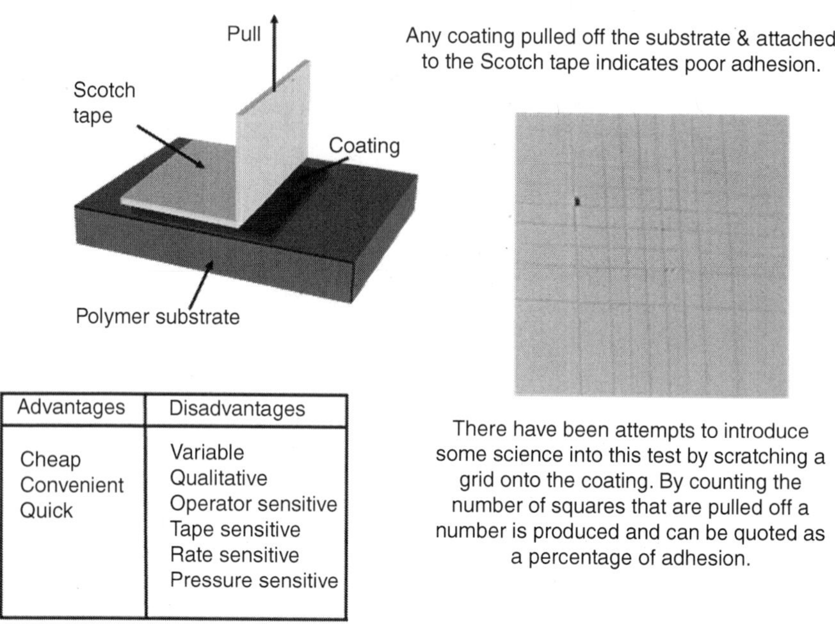

Advantages	Disadvantages
Cheap Convenient Quick	Variable Qualitative Operator sensitive Tape sensitive Rate sensitive Pressure sensitive

Figure 3.10 A schematic of the 'Scotch®' tape adhesion test.

surface and then pulled off. If any coating is pulled off at all it is an indication of poor adhesion. What this test does not tell is how good the adhesion really is. It may be that the adhesion is only marginally better than the tape adhesive or it may be considerably better than the tape adhesive strength. The problem comes in choosing what other test to use with which to test the coating adhesion. We have seen earlier that it is possible to obtain a good and accurate measure of adhesion by taking a sample of the coated barrier material and to use a micro-tensile testing system and carry out a fragmentation test but this requires time and a highly skilled operator. The same is also true of a number of the other test methods which is why the tape test ends up as being so popular despite giving such limited information.

Often tests are done that do not give any answer for adhesion other than that the vacuum coated polymer film is fit-for-purpose. There are many other adhesion tests such as scratch, scratch plus acoustic emission, blister, topple, German wheel, lap shear and peel test to name but a few. Some examples are shown in Fig. 3.11. The aim is to examine the downstream processing and final use the coated polymer film will be put to and then to chose the adhesion test that best simulates this use. Then comparing the test results with the

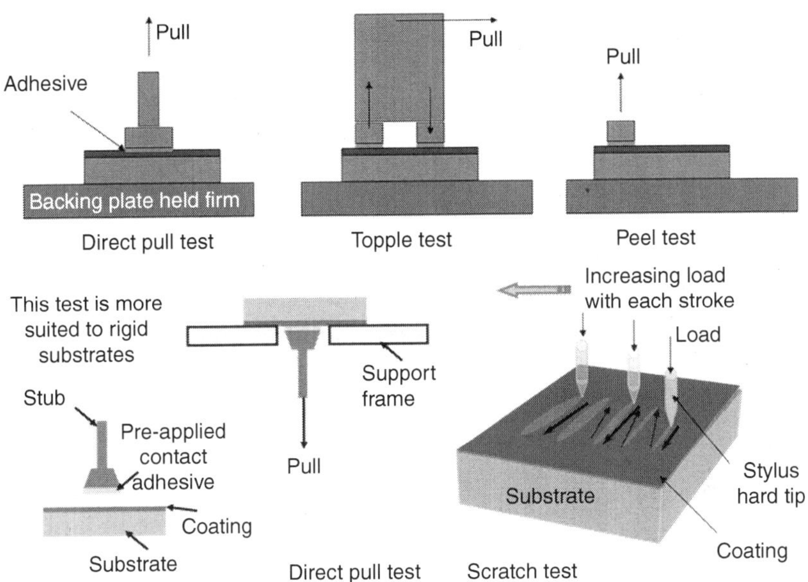

Figure 3.11 A schematic of a number of possible adhesion tests.

experience in final use will allow a correlation to be made and some level set below which the adhesion is unacceptable and above which the coated film is deemed to be fit–for-purpose. Thus the test does not have to be completed but only done to show the coated film will surpass some predetermined value.

Care needs to be taken in comparing adhesion values from different test methods as the results will often be given in different units and so cannot be compared directly. The errors and variation in the values require that many repeat tests are done of each coated sample material in order to improve the confidence of any measured value.

3.4 Pinholes

Finding and counting pinholes is very easy for metallic coatings deposited onto polymer films. The principle used is to shine a light through the vacuum coated polymer film and count the number of tiny bright spots of light. This can be done using a simple light box that has a strong light that is reflected within the box and has a translucent light diffuser for the flat surface that the coated film can be placed on. This light box provides a uniform intensity of light across

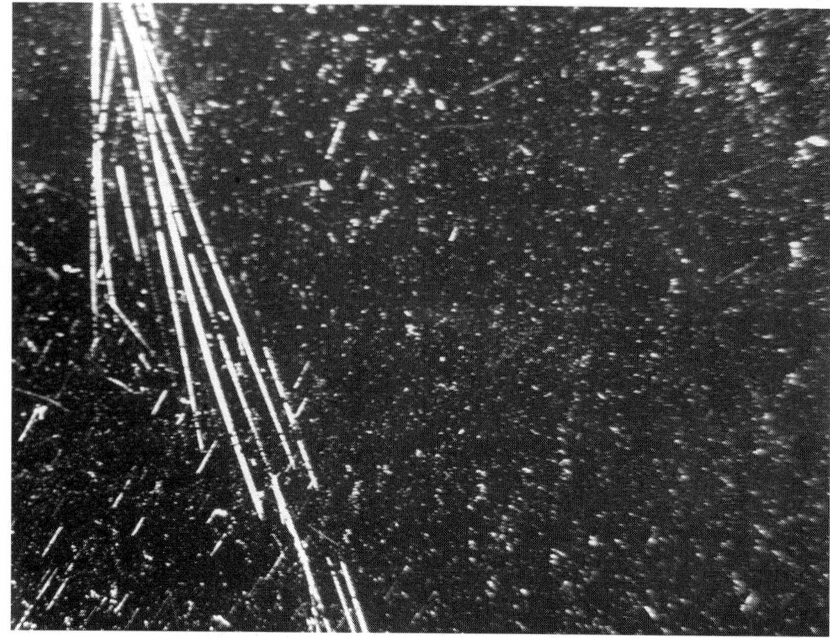

Figure 3.12 A photograph of pinholes in a metallized coated polymer film. The sample was illuminated from behind. The bright spots indicate pinholes where the light passes through unhindered. On the left hand side the coating has been scratched where the debris, when moved, has removed some of the metal leaving behind a straight line scratch.

the whole surface. This allows a large area of coated film to be examined at the same time. An example of this is shown in Fig. 3.12.

Using the same principle but using a transmission optical microscope limits the field of view but allows a more detailed examination of the pinholes and with a suitable graticule enables the diameter of the pinholes to be measured. Using a camera system this process can be automated by photographing the view and using image analysis to discriminate between dark and light areas and count and size the bright spots within each view.

As you can imagine this approach does not work for the transparent barrier coating materials. As the coatings are transparent there is no contrast and so no pinholes visible. The permeability measurements prove that they are still present as the activation energy confirms that the permeability is defect limited. Work has been done to produce a method whereby the pinholes can be highlighted in such a way that the pinholes can be counted and measured (40,41).

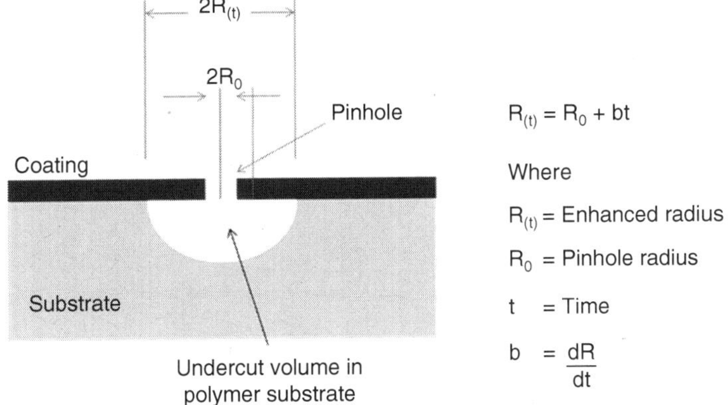

Figure 3.13 The undercutting of the polymer substrate by atomic oxygen etching to enable pinhole to be enhanced in transparent barrier coatings.

The coated material is placed in a vacuum system with the coated surface presented to an atomic oxygen plasma etching process. The etching process is much more effective in etching the polymer compared to etching the ceramic or glass-like coating and so the polymer surface that is visible at the bottom of each of the pinholes is etched away quickly. The etching is able to undermine the stiff inorganic coating and produce a volume in the polymer as shown schematically in Fig. 3.13. Following etching the coating surface can be coated with a dye and wiped clean. The areas with a complete coating will not take up any of the dye whereas the volume under the pinhole can take up sufficient quantity of dye that it immediately makes the pinholes visible.

The etch rate of the polymer will differ depending on the original pinhole size. A calibration chart can be produced where the defect size is measured after a fixed time of atomic oxygen etching. The slope of the resultant graphs if drawn back to zero etch time will indicate the original pinhole size.

3.5 Surface Energy

When we are depositing a coating there are three surface types that have energy associated with them, these are the substrate to air surface energy, the liquid to air surface tension and the substrate to liquid interfacial energy as shown schematically in Fig. 3.14. When a

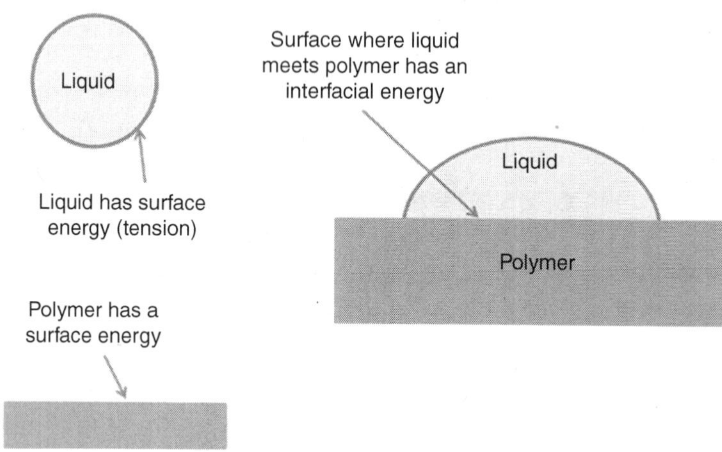

Figure 3.14 A schematic of the three types of energy that need to be minimised.

liquid comes in contact with a surface the combination of these three energies will produce a minimum total energy. It will depend on the relative energy levels as to what form the coating will take such as if the combination of the surface energy of the polymer film and the liquid surface tension is the lowest energy then the coating will form a ball on the surface. Whereas if the combination of the interfacial energy and surface tension make the minimum energy then the coating will fully wet out the surface. These extremes are shown in Fig. 3.15 along with the different conditions applied to Young's equation that would need to be satisfied to produce these results.

Measuring the surface energy of the substrate is important as it gives an indication of the potential wetting and adhesion of any coating that is deposited onto the surface. It is also used to measure the effect of any pre-treatment process to help optimise the treatment. For use on the shop floor by system operators there is a quick and easy test option where a series of dyne pens can be purchased with pens filled with liquids of differing surface tension values in increments of a couple of dynes. The pens are wiped on the substrate surface and the surface observed. If the pens liquid has a lower surface energy than the substrate surface the liquid will wet the substrate surface. This is in comparison to those pens where the liquid is of a higher surface tension than the surface energy of the substrate where the liquid on the substrate will ball up and not wet the surface. These two different conditions are shown in Fig. 3.16. The surface tension values of the pens where the wetting characteristic

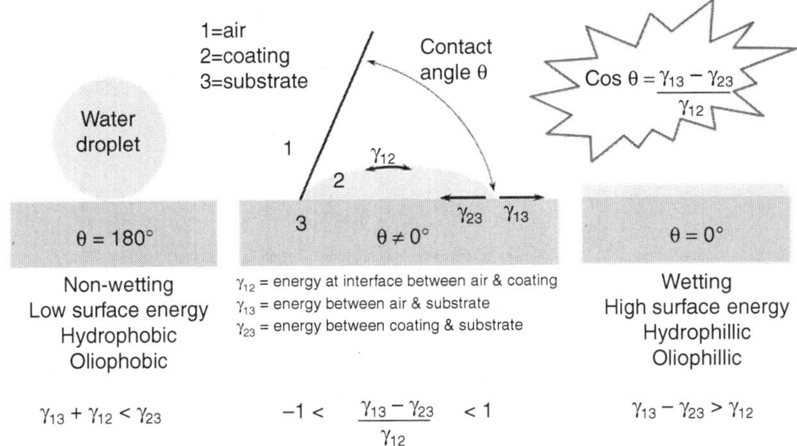

Figure 3.15 The effect on coating wetting of changing the surface energy of the substrate.

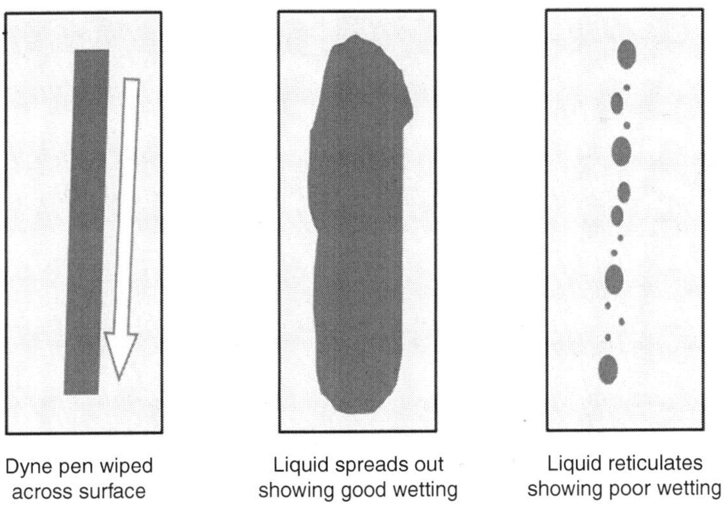

Figure 3.16 A schematic of what to look for when using dyne pens.

of the liquid on the substrate changes from non-wetting to wetting will give a measure of the substrate surface energy (42).

The other methods of measuring the surface energy are somewhat more time consuming and require some test equipment and a higher level of operator skill (43). A drop of water is measured out using a syringe and dropped onto the surface of the substrate to be measured (44). The water will form a drop on the surface where there is a contact area between the water drop and the substrate. The larger the contact surface area that is formed between the water and the substrate the higher the substrate surface energy and the better the wetting of the substrate (45–47). The same volume of water will thus form a different contact angle between the liquid and the substrate when the substrate surface energy is low and the water non-wetting and when the substrate surface energy is high and the water wets the surface well. This contact angle can be measured by projecting a light from one edge of the polymer web so that an image of the web and water drop is produced on a screen allowing the angle to be measured by a protractor. With modern machines this can now be imaged and put onto a computer and the angle defined and the surface energy calculated automatically.

The higher the substrate surface energy the more the coating spreads out and in the case of vacuum deposited coatings the coating will become continuous at a thinner coating thickness than for a coating deposited onto a lower surface energy substrate. The lower the surface energy the higher the individual coating islands will become and so the coating will end up with more pores and with a thicker coating required to produce a continuous coating.

Although it is possible to use the various forms of plasma treatment (flame, corona, atmospheric plasma, vacuum plasma) to raise the surface energy of polymer films the process needs to be used with care as it is possible to over treat the surface and to reduce the adhesion. This is shown schematically in Fig. 3.17. Initially as the plasma treatment is increased the surface energy increases. If this is continued the surface energy will achieve a maximum value where the measure of surface energy will plateau out. If the adhesion is plotted out we can see that there will be a similar initial improvement and rise to a maximum but instead of the adhesion remaining at a plateau it falls back. If the treatment is continued the adhesion may even reduce to an adhesion less than the original level.

The other aspect of surface treatment and surface energy that also needs to be considered is that of the ease of handling the coated barrier film and the back surface of the film substrate. Treating the polymer surface to clean it and also raise the surface energy can also result in the coefficient of friction being increased too. The higher the coefficient of friction the more problems in handling the polymer

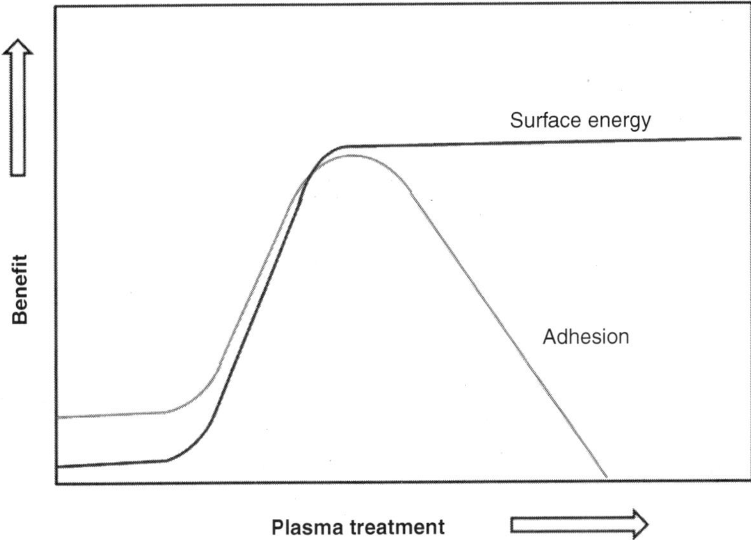

Figure 3.17 A graph of the trends of surface energy and adhesion.

film in downstream packaging operations. Often the inclusion of fillers into the substrate are to increase the surface roughness and in so doing reduce the contact area and also the coefficient of friction which in turn makes the handling easier. The other common method of reducing the coefficient of friction is to add slip agents to further reduce the coefficient of friction. Both of these methods of making the handling better are detrimental to producing barrier coatings with minimal defects. The slip agents tend to reduce the surface energy of the substrate which stops the vacuum deposited coating from wetting the surface well and any filler protruding from the surface will be a source of poor coating quality and cracks and hence poor barrier performance. It is common to pre-treat the polymer film surface that is to be vacuum coated in order to increase the surface energy and hence adhesion. With the protruding fillers it is necessary to add a coating to cover up the asperities and produce a new smoother surface to reduce the defects and improve the barrier performance. What often gets neglected is what happens to the back surface of the polymer film substrate. The back surface may have filler protruding from the surface and, if slip agents are used, there will be low surface energy contaminants present. These can damage the vacuum coated barrier coating. The fillers can scratch the vacuum coating, or if the adhesion is not sufficient the fillers can cause coating pick-off where, following re-winding, the filler is pressed so hard against the coating it is possible for the coating to stick to the filler in preference to sticking on the polymer substrate. If this occurs the coating is picked off the next time the film is unwound. If the back surface is also treated to clean it up of low surface energy material it can end up that the film surfaces will no longer slide across each other as easily and so it becomes harder to wind the roll up well. Also the stick-slip movement of the surfaces against each other can result in more damage to the vacuum deposited coating from any protruding filler. If there is slip agent present on the back surface this low surface energy material, once in close contact with the freshly deposited coating, which will be at a very high surface energy (cf. aluminium at 840 dynes/cm), will be transferred across to the vacuum coated surface reducing this surface energy down towards matching the surface energy of the back of the film. This means that any laminating layer or additional coating that needs to be added to the vacuum coated layer will potentially have problems with adhesion. It will certainly require the vacuum deposited coating to be treated to remove any of this low surface energy contaminant.

This means that those doing the vacuum coating can end in a no-win situation. If they clean both surfaces to maximise the adhesion and keep the coated surface clean, for further downstream coatings or lamination, then they will produce a film that can have winding problems that may result in damaging the vacuum deposited high barrier coating and reducing the final barrier performance. If they do not clean the surface the winding may be easier but getting good adhesion may be problematic and may require an additional process step of cleaning and raising the surface energy of the vacuum deposited coating.

3.6 Coefficient of Friction

It is useful to monitor the coefficient of friction of the substrate and coated substrate. Used in collaboration with the measurement of surface energy the coefficient of friction can help in comparing substrate properties relating to handling and adhesion problems.

There are two measures of friction the static and kinetic friction. Generally the friction is measured between a film sample sliding over itself, although it can be over some other defined material. The

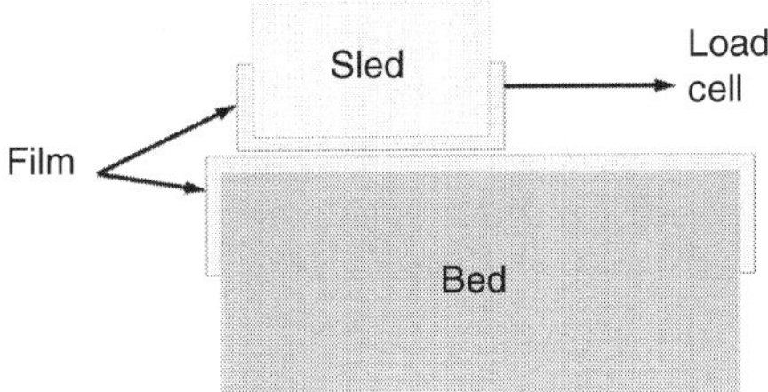

Figure 3.18 A schematic a Coefficient of Friction (CoF) test.

static friction is a measure of the force required to get the sample moving over the surface. The kinetic friction is the force required to keep the movement going at a predetermined speed. A schematic of the test is shown in Fig. 3.18.

Both the static and kinetic frictional measurements can vary with time. Many films have a problem with blocking and this can affect the static friction. The longer the sample sits on the test bed before the test begins the more the sample has time to block and the rate of application of the load also has a large effect on the static friction measurement. The age of the sample can also have an effect as exudates in the samples, such as oligomers or slip agents, will appear on the surface, a process known as blooming, and these will affect the surface energy and friction. The rate of migration of the oligomers or slip agents depends on the molecular size of them, the crystallinity of the polymer and the temperature. The shorter the chain length of the low molecular weight slip agent, the more amorphous the polymer and the higher the temperature the faster the migration rate will be. The older the sample the more exudates will appear on the surface and so the lower the friction coefficient. As the rate of blooming may vary over the surface of the film it is necessary to take repeated measurements to get a good average value. It is also important to make sure the rate of movement of the sled over the contacting surface is also kept constant in order to minimise measurement variations.

Anti-blocking fillers added to the polymer can reduce the CoF down to 0.3–0.4 which may still be higher than required to make handling easy. The shape, size distribution and type of filler can affect not only the CoF but also the surface roughness and haze of the surface. The surface roughness will also affect the barrier performance of any vacuum deposited coating. Some polymer films can be coextruded to give a smooth surface to deposit the coating onto and have a rough filled back surface. Thus the use of fillers will always be a compromise to produce the preferred balance of properties of handling, barrier, optical performance, adhesion and mechanical properties.

Slip additives may be used in conjunction with the fillers to further reduce the CoF down to around 0.2 where 0.16 – 0.2 CoF is regarded as acceptable for easy processing. As the slip agents cannot be controlled to only appear at one surface and not the other the use of these too will be a balancing act in improvements in handling but problems with adhesion.

The coefficient of friction can be measured in accordance with ASTM D1894 – 08 the 'Standard test method for static and kinetic coefficients of friction of plastic film and sheeting'. The other part of this that is of interest is the tendency of the rolls of coated film to block. Blocking is where two polymer webs once brought together cannot slip against each other and in some instances may not be easily separated. This can be made worse if the rolls have a static charge when wound up or if they are wound hot and so cool and contract adding to the compressive load on the film. The tendency to block can be measured in accordance with ASTM D3354 – 08 'Standard Test Method for Blocking Load of Plastic Film by the Parallel Plate Method' or the earlier ASTM 1893 – 67 (discontinued after 1991). Blocking is worse for smoother film surfaces, for films with high electrostatic charge and for films with a high surface energy.

3.7 Coating Thickness

Many of the vacuum deposited coatings used for barrier applications are very thin and range from 5 nm – 100 nm, and sometimes thicker (48–54). Particularly the transparent ceramic or glass-like vacuum deposited coatings are very thin, where it is believed that the thinner the coatings the more flexible they will be and so less likely to crack. With very thin vacuum deposited coatings it can be difficult to measure the thickness accurately and any error, even of only 1 nm, can be as much as 20% of the coating thickness.

A common technique used for measuring coating thickness has been to use a stylus dragged across the edge of the coating so a step height is measured as shown in Fig. 3.19. So long as there is sufficient distance travelled across the coating and then substrate so that the extrapolated levels can be confirmed as being parallel then the confidence in the thickness can be high. Where a coating is seen

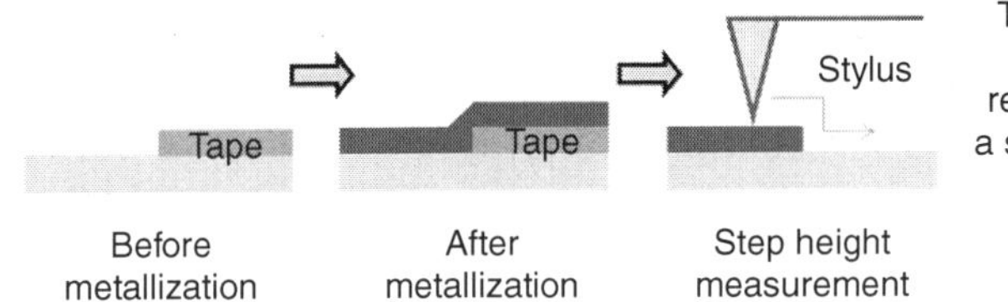

Figure 3.19 The use of tape or masking to produce a step for stylus type thickness measurements.

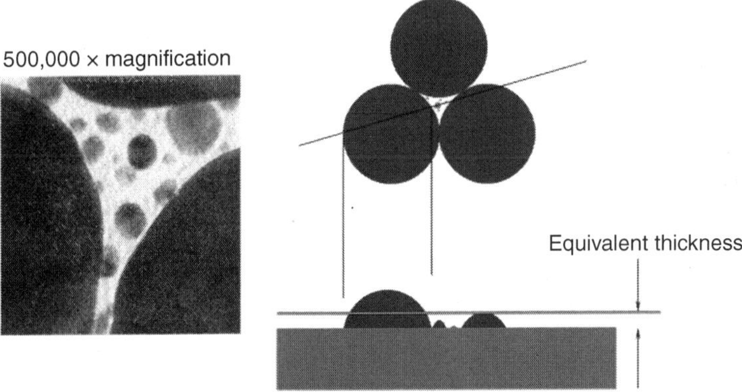

Figure 3.20 A schematic showing how very thin films can only ever have an average or equivalent thickness.

to be at a different angle and not parallel to the substrate the coating thickness then becomes much more uncertain. Not all stylus systems are sensitive enough to measure down to 10 nm or below and the ones that are able to measure down to this level need to be mounted on a vibration free surface (55,56).

Coatings where the thickness is very thin can border on the point of becoming discontinuous. Were they to be discontinuous then the barrier would be poor or none existent but at the point where the coating becomes continuous then the permeation would improve considerably. Figure 3.20 shows how the coating measured when these coatings are so thin can only reflect an average or equivalent thickness as many of the individual crystals may have nucleated at different times and have grown to different heights. As the stylus tip will be a large radius compared to the coating thickness the true variation of the surface will not be seen and so even a discontinuous coating can be measured as having an equivalent thickness. Using an atomic force microscope (AFM) does enable more of this true surface roughness to be imaged.

One of the other methods used to look at thin coatings is to pot the coated thin film in acrylic and to then shave off thin slices and look at the sample using a transmission electron microscope (TEM). An alternative to this can be to freeze the sample and fracture the frozen coated film and look at the fracture using a scanning electron microscope (SEM). Transparent ceramic or glass-like coatings have the advantage that they can usually be cracked with ease and so folding an edge can usually produce a suitable series of cracks that can provide the SEM with a suitable coating fracture edge to measure. Both of these electron microscopy techniques have to be done with some care as it is easily possible to look at the coating from an angle other than normal to the thickness and so get a false reading of the thickness. Also tilting the sample and altering the take off angle in the SEM can also change the perspective. It helps if the substrate thickness can be confirmed at the same time, although the variation in substrate thickness may only compound the measurement error. There are some calibration beads that can be added to the sample that can be used for calibration and they can help verify what is measured.

In terms of measuring the thickness of the coatings inside the vacuum system as part of the deposition process the usual method is to measure some property and to convert the measure to a thickness using a calibration chart or conversion factor. Many of these can be vacuum roll coating system sensitive. Examples of this type of measurement would be measuring the transmittance and converting it into optical density and from a calibration chart converting this to a thickness. Similarly measuring the resistance of a conducting coating and converting this measurement to a thickness is also done. All of these can result in errors in the thickness however as long as the same calibration is used consistently the conversion can be reproducible. Measuring the thickness directly can be time consuming for thin coatings and so it is common for thicker coatings to be deposited to make the thickness measurement easier. A simple method of doing this is to slow down the winding speed to increase the thickness. By slowing down the winding speed by a factor of two each time will produce an approximate doubling of the coating thickness each time.

Measuring thickness can be calculated by examining weight change between the substrate and substrate plus coating. This technique based on ASTM E252 – 06 relies on a precise knowledge of the substrate weight and area to be able to measure the weight difference. The other key factor required is to know the density of the coating. As few coatings are one hundred percent dense this density estimate can be a source of error. This change in weight can be also done by dissolving off a known area of coating. Again the coating density will affect the thickness calculated.

There are many plots of vacuum deposited coating thickness versus a variety of various other attributes (57) but often these will be from different vacuum deposition systems and may also be from different processes mixing the data to make up conversion tables can add further errors.

The electrical conductivity of the thin films depends on the density of the coating and the crystal size of the deposited material. We will see later that the nucleation and growth of the coating can be affected by the substrate and surface energy conditions as well as the quality of the vacuum system, the pressure during deposition, temperature and energy during deposition. Hence different processes and deposition systems can produce coatings with widely differing electrical conductivity at the same coating thickness (58–60).

This variation in coatings from different vacuum coating systems can lead to problems of meeting customer specifications. Specifications may be written with some measures of performance that are taken from a particular reference and this may result in a combination of properties that some other systems can never produce.

3.8 Coating Conductivity or Resistivity

Off-line the resistivity can be measured using standard four point probe measurements or using Hall effect configurations or using Van der Pauw methods (61-69) of which there are many equipment variations available.

It is more use to measure the resistivity on-line during the vacuum deposition process. It is possible to take direct measurements of resistivity of vacuum coated films by using either probes that contact the coated surface or by using conducting rolls that touch the front surface of the coating. The contact probes can be arranged in sets of four to enable four point probe measurements to be taken (70). The probes can inflict damage into the coating with smearing on ductile metal coatings and pressure cracking on brittle materials and so this technique is not preferred. The conducting rolls can have errors due to background noise as the conducting rolls need to be isolated from the rest of the system and some sliding contact needs to be made to the rolls.

To overcome these deficiencies an eddy current non-contact monitor has been developed (71–74). This uses a radio frequency (RF) supply that creates a high frequency magnetic field from a transmitting coil. The coated substrate is passed through this alternating magnetic field and if the coating is conducting there is a change in frequency and loss of signal intensity that can be detected. The change in frequency and intensity is dependent upon the coating material, thickness and the coating crystal growth morphology.

Figure 3.21 shows the two alternative methods of using this technique. On the left hand side the transmitting coil and analysing coil are on opposite sides of the coated polymer web. The measurement of resistivity is not particularly sensitive to the exact position of the coated web passing through the two coils. On the right hand side the transmitting and analysing coils are on the same side with the two coils being wound concentrically to make the measurement unit quite compact. As the magnetic field strength reduces with distance it is critical that the position of the coating to be measured is a precise distance away from the

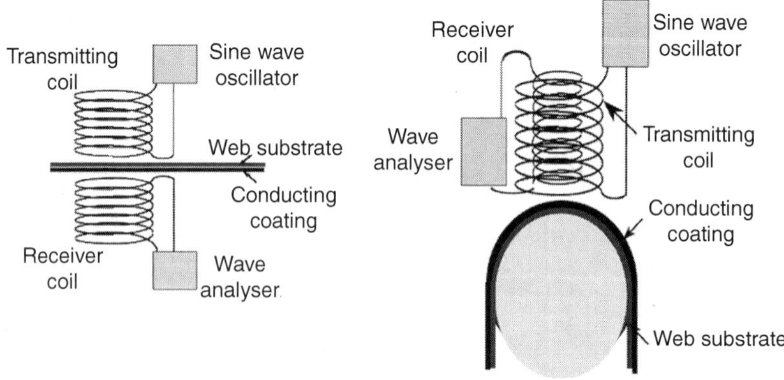

Figure 3.21 A schematic of the transmission and reflection eddy current measurement of coating resistivity.

measurement head. In order to do this the measurement is usually taken whilst the coated web is passed around a roller. This roller needs to be non-conducting otherwise the resistivity of the conducting roll will always be present as part of the measurement. This would swamp the resistivity measurement and so is eliminated by using a non-conducting roller. The measurement head can be optimised specifically for the range of coatings that it will be used on. The head will still be capable of measuring coatings with a resistivity either lower or higher than this optimum range but the sensitivity and so response will be lower than for coating within the optimum resistivity range.

As this measurement is taken over an area of the coated polymer web and the sample is moving the measurement will be an average value. Thus if this is used to convert the resistivity into a coating thickness it too will be an average value.

For on-line measurement it is usual to use an eddy current measurement head per evaporation source so that the source evaporation rate can be adjusted to bring all the measurements towards the same value.

3.9 Transmittance, Reflectance and Ellipseometry

For non-conducting coatings getting an on-line measurement can be more difficult. Transmittance or reflectance measurements can be taken and compared to a known standard to make sure the coating is of the right colour and not absorbing. The measurement of transmittance along with reflectance allows the absorptance to be calculated. The addition of transmittance plus reflectance plus absorptance equals 1. Silica coatings have a lower barrier performance when they are stoichiometric and it is often preferred to produce a substoichiometric oxide. However if the oxygen deficiency is too much there is an absorption at the blue end of the spectrum and the coatings appear yellow. It is possible to use a scanning spectrophotometer to measure the whole of the visible spectrum of transmittance for the coating. Alternatively it is possible to use a series of light emitting diodes (LEDs) of different wavelengths to monitor the critical areas of the spectrum. This can be a cheaper method of monitoring specific coatings (75).

With thick coatings the interference can give information about the thickness of the coatings but as there are no interference fringes on very thin coatings this is no help in measuring the thickness of these coatings.

One alternative technique is to use ellipsometry where elliptically polarised laser light is reflected off the coated substrate and the rotation and attenuation of the polarised light is used to measure the refractive index and thickness of the coating (76,77). This technique only works well for non-absorbing coatings. The refractive index of the substrate needs to be known and it is important to make sure there is a minimised contribution from any reflection from the back surface of the substrate. It is important that the coated surface is in a stable position and so this too is a measurement that is best taken whilst the coated polymer web is wrapped around a roller. This roller may also be blackened to help minimise any reflection from the back surface. This method is available but is not commonly used.

There is an argument that states that if you have to measure the coating thickness it is because the deposition process is not well enough under control. If the deposition parameters are well enough controlled and are stable the coating should be predictable and reproducible. This is the case for magnetron sputtering where many processes are controlled simply by the voltage, current and deposition pressure or reactive gas partial pressure. Evaporation is more difficult as very small fluctuations in temperature can give rise to much higher changes in evaporation rate. Thus it is more important for evaporation processes to have downstream monitoring to provide feedback information that can be used to control the process more accurately.

3.10 Standard Test Methods

3.10.1 Permeability Tests

ASTM D3985 - 05 Standard test method for oxygen gas transmission rate through plastic film and sheeting using a coulometric sensor.

ASTM F1927 - 07 Standard test method for determination of oxygen gas transmission rate, permeability and permeance at controlled relative humidity through barrier materials using a coulometric detector.

ASTM F1307 - 02(2007) Standard test method for oxygen transmission rate through dry packages using a coulometric sensor.

Also DIN 53380-3 or JIS K 7126.

ASTM D1434 - 82(2009)e1 Standard test method for determining gas permeability characteristics of plastic film and sheeting.
Also DIN 53380-1.

ASTM F1249 - 06 Standard test method for water vapour transmission rate through plastic film and sheeting using a modulated infrared sensor.
Also BS 7406 method A, ISO 9932 or TAPPI T557.

ASTM E96 / E96M - 05 Standard test methods for water vapour transmission of materials.
Also BS 2782:820A, BS 3177, DIN 53122 and JIS Z 0208

3.10.2 Other Mechanical or Optical Performance Tests

ASTM F392 - 93(2004) Standard test method for flex durability of flexible barrier materials.

ASTM D2578 - 09 Standard test method for wetting tension of polyethylene and polypropylene films.

ASTM D5946 – 96 Standard test method for corona-treated polymer films using water contact angle measurements.

ASTM D3359 - 09 Standard test methods for measuring adhesion by tape test.

ASTM D1894 - 08 Standard test method for static and kinetic coefficients of friction of plastic film and sheeting. (not equivalent to ISO 8295–1995).

ASTM D1893 – 67 Earlier standard for measuring blocking of polymer films. Discontinued in 1991 but still often quoted.

ASTM D5534 – 08 Standard Test Method for Blocking Load of Plastic Film by the Parallel Plate Method. (Test method is similar to ISO 11502 Method B, but not equivalent)

ASTM D3363 - 05 Standard test method for film hardness by pencil test.

ASTM D4060 - 07 Standard test method for abrasion resistance of organic coatings by the Taber Abraser.

ASTM D1044 – 94 Standard test method for resistance of transparent plastics to surface abrasion.

ASTM D522 - 93a(2008) Standard test methods for mandrel bend test of attached organic coatings.

ASTM B659 – 97 Standard guide for measuring thickness of metallic and inorganic coatings.

ASTM D-3002-71 Evaluation of Coatings for Plastics.

ASTM D1003 – 97 Standard test method for haze and luminous transmittance of transparent plastics.

ASTM E252 - 06 Standard Test Method for Thickness of Foil, Thin Sheet, and Film by Mass Measurement

References

1. ASTM D – 3985-95 Standard test method for oxygen gas transmission rate through plastic film & sheeting using a Coulemetric sensor. American Society for Testing & Material. Published 1995.
2. ASTM – F1249- 06 Standard test method for water vapour transmission rate through plastic film and sheeting using a modulated infrared sensor. American Society for Testing & Material. Published 2006.
3. Hartvigsen A. 'Automatic Permeability Testing: The Challenges and Solutions'. 48th Ann. Tech. Conf. Proc. Society of Vacuum Coaters 2005, pp 184–188.
4. Stevens M. et al. 'Water vapor permeation testing of ultra-barriers: limitations of current methods and advancements resulting in increased sensitivity'. 48th Ann. Tech. Conf. Proc. Society of Vacuum Coaters 2005, pp 189–191.
5. Oxtran Operating Manual. Modern Controls (Mocon) Inc.
6. Nörenberg H. et al. 'Mass spectrometric estimation of gas permeation coefficients for thin polymer membranes'. *Rev. Sci. Instruments*, 70, 1999, pp 2414–2420.
7. Nörenberg H. et al. 'A new method for measuring low levels of water vapour permeation through polymers and permeation barrier

coatings'. 45th Ann. Tech. Conf. Proc. Society of Vacuum Coaters 2002, pp 546–547.
8. Nörenberg H. 'Outgassing and permeation studies of polymer substrates for barrier films'. 48th Ann. Tech. Conf. Proc. Society of Vacuum Coaters 2005, pp 631–633.
9. Nörenberg H. et al. 'Permeation of gases through polymer membranes investigated by mass spectroscopy'. Vacuum 53, 1999, pp 313–315.
10. G. Nisato G. et al. 'Evaluating high performance diffusion barriers: the calcium test'. 21st Ann. Asia Display, 8th Internat. Display Workshop, Nagoya, Japan, Oct 2001.
11. Kumar R.S. et al. 'Low moisture permeation measurement through polymer substrates for organic light emitting devices'. *Thin Solid Films*, Vol. 417, No 1–2, 2002, pp 120–126.
12. Paetzold R. et al. 'Permeation rate measurements by electrical analysis of calcium corrosion'. *Rev. Sci. Inst.*, 74, 12, 2003, pp 5147–5150.
13. Paetzold R. et al. 'High-sensitivity permeation measurements on flexible OLED substrates'. Proc. SPIE 5214, Organic Light-Emitting Materials and Devices VII, 2004, pp 73–82.
14. Louch S. et al. 'The calcium test as a tool for evaluating the performance of flexible barrier films'. 51st Ann. Tech. Conf. Proc. Society of Vacuum Coaters, 2008, pp. 803–807.
15. Louch S. et al. 'Efficacy of flexible moisture barrier films produced using a roll-to-roll coater as measured by the calcium test'. 52nd Ann. Tech. Conf. Proc. Society Vacuum Coatings 2009, pp 760–764.
16. Carcia P.F. et al. 'Ca test of Al2O3 gas diffusion barriers grown by atomic layer deposition on polymers'. *Applied Physics Letters* 89, 031915 (2006), pp 1–3.
17. Otsuka M. et al. 'Study of transparent high gas barrier film and the evaluation method of water vapour transmission rate (WVTR)'. 51st Ann. Tech. Conf. Proc. Society of Vacuum Coaters, 2008, pp. 814–817.
18. Ubrig J. et al. 'Interface phenomena in multilayers deposited by PECVD for encapsulation of lithium microbatteries'. *Journal of Physics:* Conference Series 100 (2008) 082030 IVC-17/ICSS-13 and ICN+T2007 pp 1–4.
19. Dunkel R. et al. 'Method of measuring ultralow water vapour permeation for OLED Displays' Proc. IEEE, 93, 8, 2005, pp 1478–1482.
20. Dunkel R. et al. 'A new method for measuring ultra-low water-vapor permeation for OLED displays' *R. J. Society of Information Displays* 13/7, 2005, pp 569–574.
21. Groner M.D. et al. 'Gas diffusion barriers on polymers using Al2O3 atomic layer deposition' *Applied Physics Letters* 88, Art No. 051907 2006, pp 1–3.

22. Groner M.D. et al. 'Gas diffusion barriers on polymers using Al_2O_3 atomic layer deposition' 48th Annual Technical Conference Proceedings (2005) pp 169–172.
23. Jameson E.H.H. *'The structure and barrier properties of metallized polyester film'* PhD Thesis. Cambridge Univ. UK 3rd Feb 1981.
24. Jamieson E.H.H. & Windle A.H. 'Structure & oxygen-barrier properties of metallized polymer films' *J. Matls. Sci.* 18, 1983 pp 64–80.
25. Bouten P.C.P. 'Failure test for brittle conductive layers on flexible display substrates' Proc. EuroDisplay 2002, paper 17-5, pp 313–316.
26. Specht J. 'Metalization: An End-User's Perspective' 41st Ann. Tech. Conf. Proc. Society of Vacuum Coaters 1998, pp 440–445.
27. Hoekstra T. 'Metallized materials for FMCG packaging trends and needs' Proc. 3rd Ann. Vac. Coating & Metallizing Conf. 2004, Section II.
28. Stoney G.G. 'The tension of metallic films deposited by electrolysis' Proc. R. Soc. Lond. A 82, 1909, pp 172–175.
29. Chow T.S. et al. 'Direct determination of interfacial energy between brittle and polymeric films. *J. of Polymer Sci: Polymer Physics Edn.* Vol. 14, Issue 7, 1976, pp 1305–1310.
30. Rochat G. et al. 'Mechanical analysis of ultrathin oxide coatings on polymer substrates in situ in a scanning electron microscope' *Thin Solid Films* 437 (2003) pp 204–210.
31. Leterrier Y. and Månson J.-A. E. 'Mechanical integrity of thin films on polymer substrates' Proc. AIMCAL Fall Tech Conf. 2004.
32. Marras L. and Sbaizero O. 'The fragmentation test applied to adhesion measurements and microstructural characterisation in plasma pretreated metallized plastic webs' *Packaging Technology and Science* Vol 22, No. 5, 2009 pp 293–302.
33. Wheeler D.R. and Osaki H. 'Intrinsic bond strength of metal films on polymer substrates - A new method of measurement' ACS Symposium Series, Vol. 440 *'Metallization of Polymers'* Chapter 36, 1990, pp 500–512.
34. Leterrier y. et al. 'Layer mechanics - Experimental methods and models' D4PU FLEXled-epfl-0209-002/Leterrier Sept. 2002, IST-2001-34215.
35. Rochat G. et al. 'Mechanical properties of SiOx coated PET films characterised by AFM equipped with micro-tensile stage' Proc. AIMCAL Fall Tech Conf. 2007.
36. Leterrier Y. 'Durability of Nanosized Oxygen-Barrier Coatings on Polymers' Prog. *Mater. Sci.*, 48, (2003), pp 1–55.
37. Felts J.T. 'Transparent Barrier Coatings Update: Flexible Substrates' 36th Ann. Tech. Conf. Proc. Society of Vacuum Coaters 1993, pp 324–331.
38. Howells D. et al. 'The influence of the polyester substrate on the structure and performance of vacuum-deposited coatings' 48th Ann. Tech. Conf. Proc. Society of Vacuum Coaters 2005, pp 638–643.
39. Fayet P. et al. 'Effect of anti-blocking particles on oxygen transmission rate of SiOx barrier coatings deposited by PECVD on PET films' 48th Ann. Tech. Conf. Proc. Society of Vacuum Coaters 2005, pp 237–240.

40. da Silva Sobrinho A.S. et al. 'Detection and characterization of defects in transparent barrier coatings. 42nd Ann. Tech. Conf. Proc. Society of Vacuum Coatings 1999, pp 316–319.
41. Silva Sobrinho A.S. et al. 'Defect-permeation correlation for ultrathin transparent barrier coatings on polymers' *J. Vac. Sci. Technol. A* 18.1., Jan/Feb 2000, pp 149–157.
42. ASTM 2578-99a. 'Standard test method for wetting tension of polyethylene & polypropylene films.' 1999 (or ASTM 2578-84 1984 same title) American Society for Testing & Materials.
43. Hansen C.M. 'Characterization of surfaces by spreading liquids.' *J. of Paint Technology* Vol. 42, No. 550, Nov. 1970 pp 660–664.
44. ASTM D 5946-96. 'Standard test method for corona-treated polymer films using water contact angle measurements.' American Society for Testing & Materials. Approved 1996. Published June 1996.
45. Good R.J. 'Contact angle, wetting, & adhesion: a critical review.' pp 3–36 In *'Contact angle, wetability & adhesion.'* Ed. Mittal K.L. Pub. VSP Utrecht 1993.
46. Padday J.F. 'Spreading, wetting & contact angles.' pp 97–108 In *'Contact angle, wetability & adhesion.'* Ed. Mittal K.L. Pub. VSP Utrecht 1993.
47. Wetterman B. 'Contact angles measure component cleanliness.' Precision Clean. Oct 1997 pp 21–24.
48. Fukugami N. et al. 'Density measurement of thin glass layers for gas barrier films' *J. Vac. Sci. Technol. A* 17(4), Jul/Aug 1999, 1840–1842.
49. Decker W. and Henry B. 'Basic principles of thin film barrier coatings' 45th Ann. Tech. Conf. Proc. Society of Vac. Coaters 2002, pp 492–502.
50. Phillips R.W. et al. 'Evaporated dielectric colourless films on PET and OPP exhibiting high barriers towards moisture and oxygen' 36th Ann. Tech. Conf. Proc. Society of Vac. Coaters 1993, pp 293–301.
51. Broomfield A.A. 'Uses of web coated materials in flexible packaging' 35th Ann. Tech. Conf. Proc. Society of Vac. Coaters 1992, pp 21–27.
52. Kelly R.S.A. 'Development of aluminium oxide clear barrier films' 37th Ann. Tech. Conf. Proc. Society of Vac. Coaters 1994, pp 144–148.
53. Phillips R.W. et al. US Patent 5,792,550 Aug 11th 1998 'Barrier film having high colourless transparency and method.
54. Phillips R.W. et al. US Patent 6,576,294 June 10th 2003 'Method for forming barrier films'.
55. Whitehouse D.J. 'Some ultimate limits on the measurement of surfaces using stylus techniques.' Measurement & Control Vol. 8, 1975 pp 147–151.
56. Piegari A. & Masetti E. 'Thin film thickness measurement: a comparison of various techniques.' *Thin Solid Films* Vol. 24, 1985 pp 249–257.
57. Section 2 – Electrical, optical and metal thickness relationships. AIMCAL Metallizing Technical Reference 4th Edn. 2007 Eds. Mount

III E.M. and Bishop C.A. Published by Association of Industrial metallizers, Coaters and Laminators.

58. Machlin E.S. *'Materials science in microelectromics – The relationships between thin film processing and structure'* Pub. Giro Press. 1995, ISBN 1-878857-07-X.
59. Coutts T.J. *'Electrical conduction on thin metal films'* Pub. Elsevier Scientific Publishing Co. 1974, ISBN 0-444-41184-4.
60. Zhigal'skii G.P and Jones B.K. *'The physical properties of thin metal films'* Electrocomponent Science Monographs Pub. Taylor and Francis 2003, ISBN 0-415-28390-6.
61. Valdes L.B. 'Resistivity measurements on Germanium for transistors.' Proc. IRE Vol. 42, 1954 pp 420–427.
62. American Society for Testing & Materials. ASTM F76 'Standard method for measuring Hall mobility & Hall coefficient in extrinsic semiconductor single crystals.'
63. Frank V. 'On geometrical arrangements in Hall effect measurements.' Appl. Sci. Res. B3, 1953, pp 129–140.
64. Buehler M.G. 'A hall four-point probe on thin plates.' *Solid State Electron.* Vol. 10, 1967 pp 801–812.
65. Green M.A. & Gunn M.W. 'Four-point probe Hall effect & resistivity measurements upon semiconductors.' *Solid State Electron.* Vol. 15, 1972 pp 577–585.
66. Lange J. 'Method for Hall mobility & resistivity measurements on thin layers.' *J. Appl. Phys.* Vol. 35, 1964 pp 2659–2664.
67. McAlister S. & Hurd C. 'A century of bent electrons' *New Scientist* 15th Nov 1979 pp 536–537.
68. van der Pauw L.J. 'A method of measuring specific resistivity and Hall effect of discs of arbitary shape.' Philips Res. Repts. Vol.13, 1958 pp 1–9.
69. van der Pauw L.J. 'A method of measuring specific resistivity and Hall coefficient of lamellae of arbitary shape.' Philips Tech.Rev. Vol.20, 1959 pp 220–224.
70. European Patent EU 0166477. 'Method & system for producing a reactively sputtered conducting transparent metal oxide film onto a continuous web.' 1986.
71. Zelisse J.K. 'Accurate modelling of eddy current sensors for square resistance & thickness measurements' Proc. 5th Intnl. Conf on Vac. Web Coating 1991 pp 135–143.
72. Duesbury P. 'A heuristic approach to temperature compensation in non-contact resistance meters for vacuum roll coaters.' Proc. 36th Ann. Tech. Conf. SVC 1993 pp 232–235.
73. Brandenberg J. et al. 'Experiences of operating a closed-loop feedback system on a vacuum web coater.' Proc.35th Ann.Tech.Conf. SVC 1992 pp 94–99.

74. Scarr J.M. & Zelisse J.K. 'A new topology for thickness monitoring eddy current sensors' Proc. 36th Ann. Tech. Conf. SVC 1993 pp 228–231.
75. Gale M.T. et al. 'A simple optical thin film deposition monitor using LED's and fibre optics.' *J. Vac. Sci. Tech.* Vol. 20, No. 1, 1982 pp 16–20.
76. Struempfle J. et al. 'In-situ optical measurements of transmittance & reflectance by ellipseometry on glass, strips & webs in large area coating plants.' Proc 42nd Ann. Tech Conf. SVC 1999 pp 280–285.
77. Aufderheide B.E. et al. 'Methodology of multi-wavelength real-time ellipseometry control of ITO deposition on thin transparent polymer films.' Proc. 10th Vacuum Web Coating Conf. 1996 pp 106–113.

4

Materials

There are a series of materials to be considered including the polymer substrate, any coatings such as subbing or planarising layers, the barrier vacuum coating material and then any protective over-coatings.

Most polymers can be made into films and so have the potential to be used as a material for barrier applications. However the combination of price and performance has led converters to prefer certain polymers and the polymer substrates, used in vacuum deposition processing, for food packaging is dominated by polypropylene (PP) and polyethylene terephthalate (PET). These metallized or transparent barrier coated substrates may then be laminated with other polymers to provide some other property that is not available from the basic coated substrate.

Ecological considerations have led to the development of a variety of other polymer options with biodegradable, recyclable or compostable options becoming available. The cost and performance of these polymers means that currently these are used in very small quantities although the volumes are growing. It would appear as if metallized polylactic acid (PLA) films is leading the field at present in the 'green' coated films market. Similarly there are specialist polymers such as cyclic olefin copolymers (COC) that have been developed that are also increasing in use. The COCs are used as blends, co-extrusions or as film.

In laminations nylon and polyethylene (PE) films are common additions to the vacuum deposited barrier PET or PP. Occasionally nylon or PE films are coated directly.

Figure 4.1 shows a variety of polymers and the relative barrier performance. These values are approximate but do show that uncoated they are not good enough to meet the requirements for many of the food packaging uses. Figure 4.2 shows how PET and

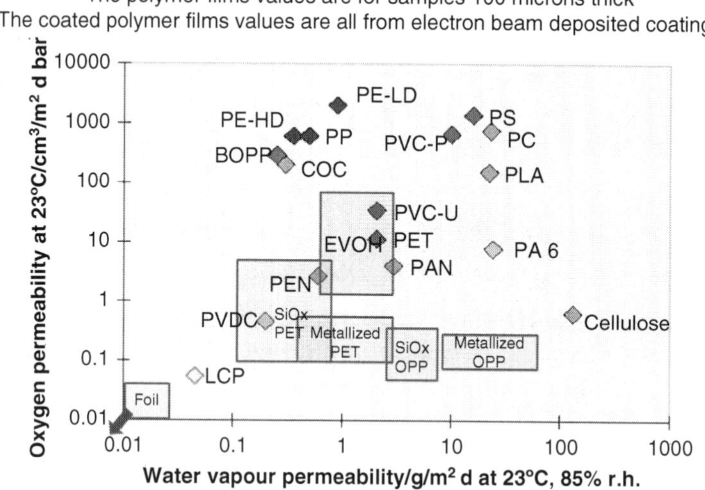

Figure 4.1 An indication of the relative barrier performance of some polymer films.

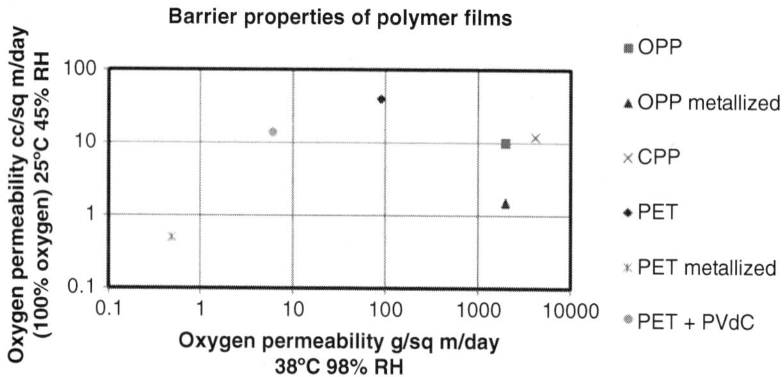

Figure 4.2 A plot showing the benefits of metallizing OPP and PET.

PP can be improved by the biaxial orientation process and further improved by the vacuum metallizing process.

Even with the two main polymers used in vacuum coating machines there are a number of different film options that provide substrates with different properties and costs.

The polymer may be used in its cast form or may be uni-axially or bi-axially oriented. The orientation can change the degree of crystallinity of the polymer and this can change the barrier performance of the polymer substrate with the higher the degree of crystallinity providing improved barrier (1). The polymer can be unfilled or filled and where filled the level, composition and size distribution of filler can be varied. The polymer can be a homopolymer or can be co-extruded so that a filled polymer can have an unfilled smoother layer co-extruded on one side. The films can also be produced at a wide variety of different thickness options. The trend in food packaging is to down-gauge. Down-gauging is where the aim is to use a thinner grade of polymer substrate but retain all the performance of the original material. Not only does this reduce the material costs but also companies may use the down-gauging as a part of an energy saving programme.

Figure 4.3 shows one example of down gauging. The first target was to eliminate the aluminium foil from the laminate. This was replaced with a metallized PET with some additional barrier performance being gained by using a nylon film (OPA). It was found that

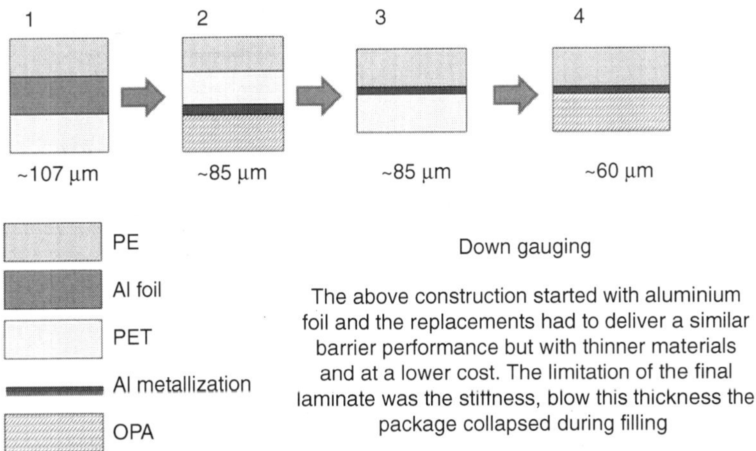

Figure 4.3 A schematic of the down gauging of a packaging material.

the package performance could be maintained without the polyethylene layer by increasing the thickness of the other two polymer layers. Finally by replacing the metallized PET with metallized OPA a further reduction of the polymer films was possible. It was also found that the barrier performance could be maintained with thinner polymers but the package was unacceptable because of handling problems. The packaging film was formed into a rectangular shape before filling and during the filling operation the package kept collapsing with the thinner polymer films. Hence the limitation in this particular case was due to the mechanical performance of the completed structure and not the barrier performance.

Specifying the substrate can be difficult as it is easy to specify the roll length, width and thickness but is much harder to specify the surface quality. The surface energy can be specified to be above a particular value but as this may change with time it should also give some indication of surface energy level within a period of time of delivery and having been stored in defined conditions of temperature and humidity. What may not be easily demanded within the specification are the surface roughness and the contamination level. I have seen in a specification the phrase the substrate 'must be free of wrinkles, scratches, dead-folds or any other defects which may prejudice the machinability and the quality of the final metallized product'. This phrase is open ended and could mean that no substrate would be satisfactory as all substrates will have some contamination that will lead to pinholes in the metallized layer. I have also seen this specification modified to define that the substrate contamination will be no worse than will result in a certain number of pinholes per given area in the metallized coating. This type of specification does depend on not only the quality of the substrates but also the winding and handling of the roll following metallization. What it does not help is those who are depositing transparent vacuum deposited coatings where the pinholes in the coating are not easily visible.

It will be shown later that the quality of barrier coatings is directly related to the substrate quality. Any protruding fillers, surface roughness, exudates such as oligomers, extrinsic contamination and low surface energy can result in poor adhesion, pinholes, cracking and scratches which all result in reductions in the coating barrier performance. Thus even for the lowest cost barrier coatings trying to start with the best possible substrate quality will pay dividends.

It is possible to obtain superior substrates but at a higher cost. Once the barrier applications moves from food packaging to electronic applications the preferred substrate tends to be PET as it is a more stable substrate with better mechanical performance for these applications. As the substrates undergo a substantial thermal change during the deposition process as the substrate temperature rises rapidly by as much as a hundred degrees and is then quickly quenched. This thermal change can cause winding problems and coating problems as the polymer expands and contracts and as any residual stress can cause some permanent film shrinkage. It is possible to purchase a thermally stabilised substrate where the substrate has been taken to an elevated temperature and any residual stress is allowed to disperse and shrinkage to take place so that on future excursions to the elevated temperature the film will remain more dimensionally stable. To provide further high temperature stability moving from PET to polyethylene naphthalate (PEN) can be done and this polymer too can be thermally stabilised. To make sure the surface of the polymer is as smooth and clean as possible it is possible to purchase a planarised version where a surface coating is applied to the substrate that covers any protruding fillers and many of the extrinsic particulate contaminants (2,3). This planarising layer is applied in a clean room to ensure that once applied the surface of the substrate is not recontaminated. This planarising of the substrate also effectively covers up any of the oligomer that normally exudes to the substrate surface and also is a source of coating defects and poor adhesion. The planarising layer once applied also resists the oligomer migrating through the planarising layer and so the smoothness of the planarising layer is maintained. These planarising layers as applied by the polymer film manufacturers are regarded as proprietary and few details are given regarding the thickness and chemical composition of the surface. Thus changing suppliers can result in changing the chemical composition of the surface and so may require a re-optimisation of any vacuum plasma treatment to achieve the maximum coating adhesion.

Obviously as each of these steps to improve the substrate quality is taken the substrate cost is also increasing and so a cost-benefit analysis needs to be done for each application to ascertain which improvements can be justified.

It has long been known that depositing a polymer coating before depositing the inorganic barrier coating in vacuum (4–9) has significant advantages in improving the barrier performance of the coated substrate. This polymer deposition process was originally developed for the deposition of the polymer layers in alternating polymer/aluminium multilayers for capacitors. This process was then developed for making optical multilayer coatings and in high barrier coatings. The basic process used a liquid monomer source outside the vacuum system where the monomer was metered and pumped via an ultrasonic spray nozzle into a small volume chamber with a hot plate surface. The monomer spray droplets when they hit the hot plate were flash evaporated into a vapour. The vapour was then redirected via baffles to an exit nozzle that was of uniform width across the whole substrate width. The baffles and vapour pressure within the hot vapour chamber helped make the exiting vapour uniform in flux across the substrate width. The substrate passes this polymer vapour source whilst it is held against a cooled deposition drum and this helps the vapour condense on the polymer web. The substrate with the monomer coating is then passed past an electron beam curtain that cures the monomer into the polymer. This polymer deposition process has to be suitably controlled if the down-the-web uniformity is to be achieved. Small variations in the monomer liquid pressure or flow through the metering pump will be exaggerated as the liquid is converted to a vapour inside the vacuum system. An alternative of evaporating a monomer directly onto the polymer substrate has been evaluated and is believed to be possible but the accuracy of the evaporation source temperature control and source temperature uniformity are critical factors if a reproducible coating thickness is to be possible (10,11). Similarly in-vacuum printing of the monomer is also believed to be possible as it is not dissimilar to the printing of oils as used in pattern metallization (12).

There are many printing processes that potentially can be used for coating the polymer substrate with a monomer (13,14). The aim is to coat the substrate with a coating that is thick enough to cover the surface defects such that a new flat and smooth surface is produced. If deposition processes such as gravure coating is used it is essential that the substrate surface energy is controlled and that suitable time is given so that the liquid surface can flow and produce the flattest surface possible (15–18). Not only does the liquid require time for levelling but it also requires time during curing to allow for the loss of material as the solvent component evaporates. If this is not done then the coating will end up with mud cracks. This subbing or planarising layer is usually of the order 500 nm or greater in thickness. Polymer substrates that have not been cleaned will have particles on the surface much larger than this and so the coating would not fully hide these particulate contaminates thus it is preferable to clean the polymer surface before adding the polymer coating. Cleaning using tacky rolls (19) can remove all particles above a particle size of approximately 300 nm which would mean the residual particles would be smaller than the coating thickness and hence would not disturb the smooth polymer surface.

Depositing polymers inside the vacuum system has been developed mostly using acrylates. When the liquid monomer is pumped into the vacuum system it is important that the monomer does not fractionate during flash evaporation nor does it polymerise or degrade on the hot plate rather than evaporate into a vapour. This limits the chemistry that can be used by this technique. Coating outside the vacuum system allows many more polymer types to be used. It has always been perceived that depositing the polymer in the vacuum system has the advantage that there is a minimum amount of time in which the new polymer surface can be contaminated and hence this will give the greatest improvement. However many groups are successfully producing ultra barrier performance materials by purchasing already planarised substrates ready for coating.

It has been shown (20) that not only is it important to have a suitably perfect surface on which to deposit the vacuum deposited inorganic coatings but it is similarly important to protect the coating from damage. Damage to the inorganic coating is most easily done if there are still contaminant particles left on the surface which can be rolled or skidded along the coated surface leaving behind a pinhole and creating a scratch. However even without these contaminant particles there can be damage done by anything hard that contacts and slides across the surface. This includes filler particles protruding from the back surface of the polymer substrate. The back surface will be in hard contact with the freshly deposited inorganic coating as the polymer web is re-wound. As the re-winding is in vacuum there is no interleaving air to help separate the layers and so even without much tension on the web the rolls will still be wound up hard. If there is any slip between layers this can cause damage to the inorganic coating. Therefore if a protective coating can be added on top of the inorganic coating it will prevent this type of damage. The in-vacuum inorganic deposition process makes coating before and after the inorganic layer deposition possible. Ideally this would be done as a single pass to prevent any possibility of damage as the web is re-wound. As this layer is (hopefully) not trying to cover up any particles but is only protecting the inorganic coating it need not be as thick as the initial planarising layer. This polymeric protective coating has the additional benefit that if there are any defects in the inorganic coating the polymer will wick down into the gaps, pores or pinholes and the diffusion rate of gas or water vapour in the polymer is several orders of magnitude lower than it would be if the holes were not filled with polymer but filled with air.

Few substrates are perfect even after the addition of a planarising layer and so it is more than likely that there will be defects in the inorganic coating which means that the barrier performance will also not be perfect but will be limited. If an ultra barrier performance is required it may be necessary to add layers. This may require the substrate being re-wound in the vacuum system and passed across the deposition sources a number of times.

Figure 4.4 shows schematically a number of the different options to make improvements to the uncoated polymer film. There is not a single way of achieving these improvements. One route is to try for perfection (21,22). In one case a smoothed PEN polymer film is coated with an ion assisted alumina film that forms a very dense amorphous coating. Then follows a similar idea but uses an aluminosilicate coating that is of the order 200 nm thick. This is quite a thick coating and is different to many that use very thin glass-like or ceramic coatings in order to get the flexibility. The other approach for the ultra barrier performance is to use multiple alternating organic and inorganic layers (23,24).

The coating materials tend to be fairly limited for food packaging with the opaque packaging being almost exclusively aluminium

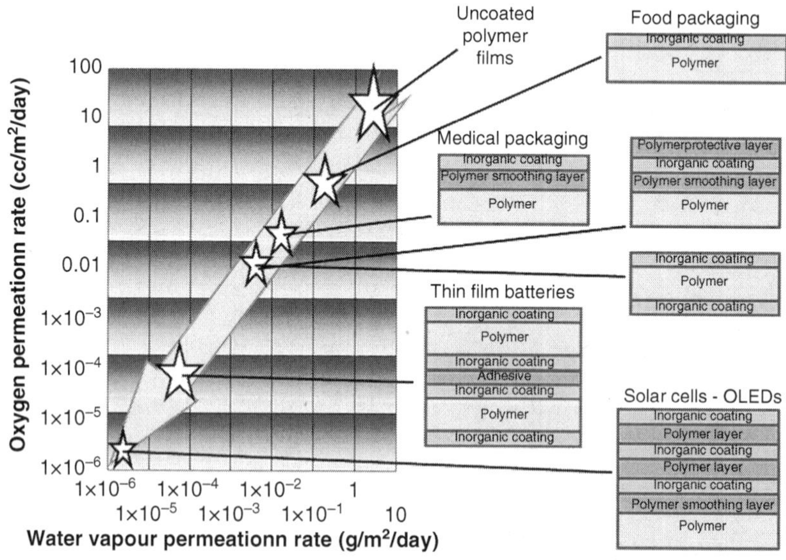

Figure 4.4 A schematic of the progression coating additions necessary to deliver the orders of magnitude improvements required to convert uncoated polymer films into ultra barrier films.

and the transparent packaging materials being shared between aluminium oxide and a sub-stoichiometric silica. There are small quantities of other materials that are trying to break into the high volume packaging markets such as hydrogenated diamond-like carbon (25,26), melamine (27), mixed oxides (28–30) or proprietary coatings (31). Proprietary coatings are often used for a captive market and often are later to be found to be variants of existing coatings. These alternative materials have remained small volumes for a number of years and do not look as if they will displace the dominance of the alumina or silica based products.

As the barrier requirements for the transparent barrier materials increases the materials that have been investigated increases too. Work done has suggested that alumina and silica are not necessarily the best barrier materials (32) and their early results suggested that some materials such as aluminium oxynitrides were no better (33) but sputtered indium tin oxide was considerably better. Since then other materials have been investigated (34) such as silicon nitride, titania and zinc tin oxide. There was some work done on depositing a mixture of silica and a polymer for space applications where the polymer was expected to help the silica withstand cracking as the material was thermally cycled between night time and day time temperatures on the outside of space craft equipment. This idea of incorporating an organic content has been incorporated by some by adding some hydrocarbon gas content during the deposition process (35). A similar process of adding a polysiloxane has also been developed (36). This grading of the material and including an organic component to the material makes the coating both more robust as well as improving the basic barrier performance and adhesion (37,38).

As nobody has deposited all the different materials by exactly the same deposition process and onto the same substrates it is safe to say that some of the information suggesting that some materials are better than others is misleading. The use of plasma enhanced chemical vapour deposition as opposed to magnetron sputtering will have very different plasma conditions and will affect the nucleation and growth of the coating differently resulting in different coating density and porosity. This will naturally lead to variations in the barrier performance. This is shown in Table 4.1 where silica has been deposited by PECVD and evaporation. What has been shown by the groups aiming to produce single layer ultra barrier coatings is that what is critical is the elimination of all coating defects of all types and the production of a dense vacuum deposited coating. It

Table 4.1 Details of some materials and their barrier performance.

Barrier Material	Thickness	Oxygen Transmission [ccm/m²/day/atm]	Water Vapour Transmission [g/m²/day/atm]	Deposition Process	Film type
PET Blank	12.00 µm	100	64.04		
PVDG	24.00 µm	8	0.3		
EVOH	24.00 µm	0.16–1.86 *	N/A		
m-OPA	15.00 µm	30			
Aluminized PET (single)	~ 30 nm	0.31–1.55	0.31–1.55	Evaporation	Al
Aluminized PET (double)	~ 30 nm each	0.03	N/A	Evaporation	Al
Aluminum on PE	7µm Al	0.001	N/A	Laminated	Al
SiOx on PET	10–80 nm	0.35–10**	0.46–1.24	Evaporation	SiOx
SiOx on PET	10–80 nm	0.08–1.55	0.5–5.0	PECVD	SiOx
Al_2O_3 on PET	20 nm	1.5	5.0	Evaporation, Reactive Evaporation	Al_2O_3
Al_2O_3/SiO_x on PET	50 nm	2.0–3.0	1.0	Evaporation	Al_2O_3/SiO_x
Diamond-like Carbon on PET	20 nm	2	1.50	PECVD	Diamond
Melamine on PET	12.00 µm	<5		PVD	Melamine
Melamine on OPP	20.00 µm/36nm	30		PVD	Melamine
OPP Blank	20.00 µm	1600			

*depending on relative humidity and ethylene content

**depending on process

remains to be seen if once these conditions have been satisfied if the performance of different materials can be discriminated.

What is also shown in Table 4.1 is the difference between single side and double side coated film. If we look at the barrier performance of the double side coated PET it is considerably more than double the performance of the single side coated film. Experiments have been done where the single side coating is twice the thickness of each of the coatings on the double side coatings. This was done to eliminate the metal thickness from the extra barrier improvement. The additional improvement is due to the pinholes in each of the coatings being sufficiently offset so that there is an increase diffusion path through the polymer which changes the polymer effective thickness as described in chapter 2.

Where the vacuum deposited coatings are less than perfect it has been shown that filling the pores and pinholes with something other than air has the advantage of reducing the diffusion coefficient in these defects. What is more, if the polymer can be modified by the addition of a filler to have an even lower diffusion coefficient than the unfilled polymer the blocking up of the defect holes can be further improved (39). This takes the use of large aspect ratio clay flake fillers that have been added to polymers and thin polymer coatings (40–44) down to the nano scale where nanoflakes of a similar aspect ratio are added to the polymer so that the flakes settle within the pinholes or pores and this increases the tortuous path of any diffusing gas or water vapour and so improves the barrier.

4.1 Packaging Materials Calculations

In deciding what materials to use, or to check if a particular combination of materials will deliver the required barrier performance, it can be useful to calculate what these parameters might be. Below are a couple of examples that might prove useful.

If we use for our example a package of a surface area of 500 sq cm which contains 1kg of a food product that we wish to have a shelf life of 6 months (~180 days). It is determined by the manufacturers that the foodstuff can only tolerate an ingress of 10 ppm (parts per million) of oxygen before the product is deemed unacceptable.

The amount of oxygen allowed to enter the package is:

$$I_M = I_p M_p = (10 \times 10^{-6})(1000g) = 0.01g \text{ oxygen}$$

where

I_M is the maximum amount of oxygen in grams that can enter the package
I_p is the maximum allowable ingress in ppm
M_p is the mass of product.

We next need to convert this mass of oxygen into a volume.
The molar mass of oxygen is 32g/g mole
1 g-mole of an ideal gas at standard pressure and temperature will occupy a volume of 22,400 cc,
from which we obtain:

$$I_v = \frac{V_g}{m_g} \times I_m = \frac{22,400}{32} \times 0.01 = 7\text{cc}$$

where

I_V is maximum amount of oxygen in cubic centimetres that can enter the package
V_g is 22,400 cc (Volume of 1 g-mole of an ideal gas at standard conditions)
m_g is the molar mass of oxygen.

So the barrier of choice must prevent more than 7 cc of oxygen from permeating into the package over the course of 180 days.
The target transmission rate is thus:

J = 7/(0.05 sq m x 180 days) = ~ 0.78cc/sq m . day

where
J or 'flux' is the rate of diffusion of Substance A through a solid, being the amount of gas passing through a unit area of the polymer in a unit amount of time.

Thus our barrier film, whatever the construction, in its final form needs to have a transmission rate of 0.78 or better. However the reality is that the barrier film will be degraded by handling, particularly in forming the package and possibly in filling the package and so some factor needs to be used to compensate for this degradation. If, from experience of running a production line, we know that the degradation is somewhere between 2x and 4x we can include a safety factor of 5x to make sure we will always be better than this. This would reduce our required transmission rate from 0.78 to 0.78/5 = 0.156 cc / sq m /day.

If we now look at a laminate structure and what performance it might deliver.

The calculation for the total permeability can be expressed using the ideal laminate theory (ILT) as follows;

$$\frac{1}{P_{total}} = \frac{x_p/x}{p_p} + \frac{x_c/x}{p_c} + \frac{x_1/x}{p_1} + \ldots$$

where

x = total multilayer material/laminate thickness
x_p = polymer film thickness
x_c = vacuum coating thickness
x_l = laminate or other layer thickness
p_p = permeability of polymer film
p_c = permeability of coating
p_l = permeability of laminate or other layer

If we use a triple layer laminate of polyethylene (25 microns) + double layer metallized polyester (12 microns) + polyethylene (25 microns) and take the following values of oxygen permeability.

The permeability of 25 microns HDPE is approximately 3000 cc/sq m/day.

The permeability of 12 microns double side metallized PET is approximately 0.01 cc/sq m/day.

If we put these values into the above equation we get;

$$\frac{1}{P_{total}} = \frac{25/62}{3000} + \frac{12/62}{0.01} + \frac{25/62}{3000}$$

$$\frac{1}{P_{total}} = \frac{0.403}{3000} + \frac{0.194}{0.01} + \frac{0.403}{3000}$$

$$\frac{1}{P_{total}} = 0.0001343 + 19.4 + 0.0001343 = 19.4002686$$

$$P_{total} = \frac{1}{19.4002686} = 0.052\text{cc/sq m/day}$$

Note that this calculation is for demonstration only. If this were a real laminate one could expect that any adhesive used in the lamination could have a minimal effect on the calculation but it could actually improve the metallized barrier performance as the adhesive wicked into and pinholes or defects could reduce the gas permeation through the defects.

Hence any calculation should only be used as guidelines and not definitive values.

References

1. Breil J. 'Oriented film technology' Chapter 11 in *'Multilayer flexible packaging – Technology and applications for the food, personal care and over-the-counter pharmaceutical industries'* Ed. Wagner J.R. Pub. Elsevier, 2009, pp 119–136 ISBN-13: 978-0-8155-2021-4.
2. MacDonald W.A. 'Engineered films for display technologies' *J. Materials Chemistry* 14, 2004, pp 4–10.
3. Adam R. et al. 'Optimising polyester films for flexible electronic applications' Proc. AIMCAL Fall Tech. Conf 2004.
4. Affinito J.D. et al. 'A new hybrid deposition process, combining PML & PECVD, for high rate plasma polymerisation of low vapor pressure, & solid, monomer precursors.' Proc. 42nd Ann Tech. Conf. SVC 1999 pp 102–106.
5. Shaw D.M. & Langlois M.G. 'A new high speed process for vapor deposition acrylate thin films: an update.' Proc. 36th Ann Tech. Conf. SVC 1993, pp 348–352.
6. Affinito J.D. et al. 'PML/Oxide/PML barrier layer performance arising from use of UV or electron beam polymerisation of the PML Layers' *Thin Solid Films,* 19, 1997, pp 308–309.
7. Yializis A. and Mikhael M. 'UV versus electron beam radiation curing of vacuum deposited levelling coatings for high barrier applications' Proc. 47th Ann Tech. Conf. SVC 2004, pp 600–605.
8. Affinito J.D. et al. 'Vacuum deposition of polymer electrolytes on flexible substrates' Proc. 9th Internatl. Conf. Vac. Web Coating, 1995, pp 20–36.
9. Yializis A. et al. 'Ultra high barrier films' Proc. 43rd Ann Tech. Conf. SVC 2000, pp 404–407.
10. Affinito J. et al. 'Web substrate heating and a thermodynamic calculation method for Li film thickness in a thermal evaporation system' Proc. 44th Ann Tech. Conf. SVC 2001, pp 492–497.
11. Affinito J. 'Polymer film deposition by a new vacuum process. Proc. 45th Ann Tech. Conf. SVC 2002 pp 425–439.
12. Copeland N. et al. 'In-register in-vacuum pattern printing - From wish to reality' Proc. 52nd Ann Tech. Conf. SVC 2009, pp 712–715.

13. Shepherd F. *'Modern coating technology systems.'* Pub. Emap Maclaren 1995 ISBN 0-902098-18-7.
14. Brinston J.H. 'Coating & laminating' Part I Converter Oct 1992 pp 8–12 Part II Converter Nov 1992 pp 13–18.
15. Satas D. *'Coatings technology handbook.'* 1991 Pub. Marcel Dekker Inc. ISBN 0-8247-8410-3.
16. Orchard S.E. 'On surface levelling in viscous liquids & gels.' Appl. Sci. Res. Ser. A, Vol. 11, 1962 pp 51.
17. Anshus B.E. 'The levelling process in polymer powder paints – a three dimensional approach.' A. Chem. Soc. Div. Org. Coat. Plast. Chem. Pap. Vol. 33, 1973 pp 493.
18. Kistler S.F. & Schweizer P.M. Eds. *'Liquid film coating – scientific principles & their technological implications.'* Pub. Chapman & Hall 1997 ISBN 0-412-06481-2.
19. Hamilton S. 'Advances in Contact Cleaning for Yield Improvement' Proc. AIMCAL Fall Annual Technical Conference 2008.
20. Yializis A. et al. 'Barrier degradation in aluminium metallized polypropylene films' Proc. 40th Ann Tech. Conf. SVC 1997, pp 371–375.
21. Zeigler J.P. and Piner J.R. 'O_2 and H_2O barrier materials' United States Patent Application, 2004/0209126 A1 Oct 21, 2004.
22. Narasimhan M. et al. 'Dielectric barrier layer films' United States Patent No. 7,205,662 B2 Apl 17, 2007.
23. Kapoor S. et al. 'Pilot production of ultrabarrier substrate for flexible displays' Proc. 49th Ann Tech. Conf. SVC 2006, pp 625–631.
24. Harvey III T.B. et al. 'Passivated organic device having alternating layers of polymer and dielectric' United States Patent No. 5,757,126 May 26, 1998.
25. Aisenberg S. 'Diamond-like carbon deposition technology for improved barrier films' Proc. 4th Internatnl. Conf. Vac. Web Coating, 1990, pp 25–38.
26. Yoshida M. et al. 'Improvement of oxygen barrier of PET film with diamond-like carbon film by plasma-source ion implantation' *Surface and Coatings Technology* Vol. 174–175, 2003 pp 1033–1037.
27. Jahromi S. 'A new development in clear barrier coatings' Proc. 51st Ann Tech. Conf. SVC 2008, pp 799–802.
28. Phillips R.W. et al. 'Evaporated dielectric colourless films on PET and OPP exhibiting high barriers towards moisture and oxygen' Proc. 36th Ann Tech. Conf. SVC 1993, pp 293–301.
29. Phillips R.W. 'The quest for transparent moisture barrier' Proc. 6th Internatl. Conf. Vac. Web Coating 1992, pp 103–117.
30. Garcia-Ayuso G. et al. 'Effect of surface fractality on the permeability of transparent gas barrier coatings' *Adv. Mater.* Vol 9, No. 8, 1997, pp 654–658.
31. Yoshikawa M. and Oboshi T. 'The development of new transparent barrier film by the use of novel CVD process' Proc. Future-Pak 1999.

32. Henry B.M. et al. 'Permeation studies of transparent barrier coatings' Proc. 46th Ann Tech. Conf. SVC 2003, pp 600–605.
33. Erlat A.G. et al. 'Growth and characterization of transparent metal oxide and oxynitrides gas barrier films on polymer substrates' Proc. 44th Ann Tech. Conf. SVC 2001, pp 448–454.
34. Fahlteich J. et al. 'Mechanical and barrier properties of thin oxide films on flexible polymer substrates' Proc. 51st Ann Tech. Conf. SVC 2008, pp 808–813.
35. Erlat A.G. et al. 'Ultra-high barrier coatings for flexible organic electronics' Proc. 48th Ann Tech. Conf. SVC 2005, pp 116–120.
36. Fayet P. et al. 'Polysiloxane super-toughening of silica gas barrier coatings on polymer substrates' Proc. AIMCAL Tech. Fall Conf. & 23rd Internatl. Vac.Web Coating Conf. 2009.
37. Hegemann D. et al. 'Improving the adhesion of siloxane-based plasma coatings on polymers with defined wetting properties' Proc. 45th Ann Tech. Conf. SVC 2002, pp 174–178.
38. Chiba D. et al. 'Heat sterilizable gas barrier film deposited on polyamide by plasma-enhanced CVD' Proc. 49th Ann Tech. Conf. SVC 2006, pp 613–616.
39. Ramadas S. 'Nanoparticulate barrier films and gas permeation measurement techniques for thin film solar and display application' Proc. AIMCAL Tech. Fall Conf. & 23rd Internatl. Vac.Web Coating Conf, 2009.
40. Ku B-C. et al. 'Cross-linked multilayer polymer-clay nanocomposites and permeability properties' *J. of Macromol. Sci. Pt A. Pure & Appl. Chem.*Vol. 41, No. 12, 2004, pp. 1401–1410.
41. Pannirselvam M. et al. 'Oxygen barrier property of polypropylene-polyether treated clay nanocomposite' *eXPRESS Polymer Letters* Vol. 2, No. 6, 2008, 429–439.
42. Goldberg H.A. et al. 'Nanocomposite barrier coatings' Proc. AIMCAL Tech. Fall Conf. & 23rd Internatl. Vac.Web Coating Conf, 2005.
43. Brody A.L. 'Nano, Nano – Food Packaging Technology' *Food Technology*. Vol. 57, No. 12 Dec 2003, pp 52–54.
44. Sherman L.M. 'Nanocomposites – Less hype, more hard work on commercial viability' Plastics Technology – online article 2007, pp 1–4 http://www.ptonline.com/articles/200705fa2.html (1 of 4).

5

Substrates, Surfaces, Quality and Defects

5.1 Substrates

The vast majority of the roll-to-roll vacuum deposited barrier coatings are onto polypropylene and polyester webs and so we shall concentrate on these film types. These webs are most commonly, but not exclusively, extruded and biaxial oriented either using a stenter or bubble process. In broad terms the polymer thickness gives rise to the film descriptions given in Table 5.1.

The bulk of the polymer films used to produce vacuum deposited barrier coatings are in the flexible range of film thickness. Food packaging barrier materials probably have 12 micron films as the most widely coated thickness although there is a trend to down gauging where the same performance is expected using thinner substrates. Although the starting point may be a 12 micron coated film a considerable number of film products are produced by laminating one or more of these films together and so many final products are very much thicker than this.

The polymers contain regions of both crystalline and amorphous material. The crystalline material is of higher density and has a higher barrier performance than the amorphous material. If we

Table 5.1 The polymer film thickness and description.

Polymer thickness	Description
2–200 µm	flexible
200–400 µm	semi rigid, or semi flexible
>400 µm	rigid

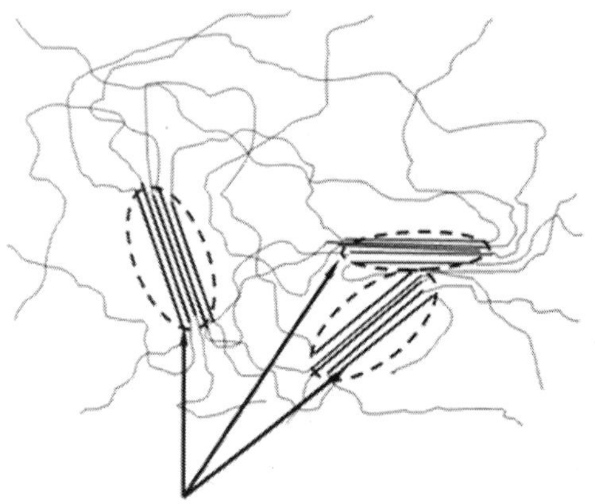

Where polymer chains run parallel crystallites can form.
The three shown here are in random orientation.
A single polymer chain may be part of several crystallites
with intermediate parts of the chain being in amorphous
regions

Figure 5.1 Cast polymer.

look at a schematic of the structure Fig. 5.1 we can see that when the polymer is first extruded as a cast film the crystalline material within the amorphous matrix are in random orientations (1). This structure changes when the polymer web is stretched and the crystalline material is oriented parallel to the web faces. The crystalline material is not only oriented but is also forced closer together. Any gas or vapour diffusing through the polymer will diffuse through the amorphous regions faster than through the crystalline regions. As it is easier for any gas or vapour to diffuse through the amorphous material the diffusion path will tend to follow a tortuous path from one surface to the other as the gas or vapour diffuses around the crystalline regions as shown in Fig. 5.2.

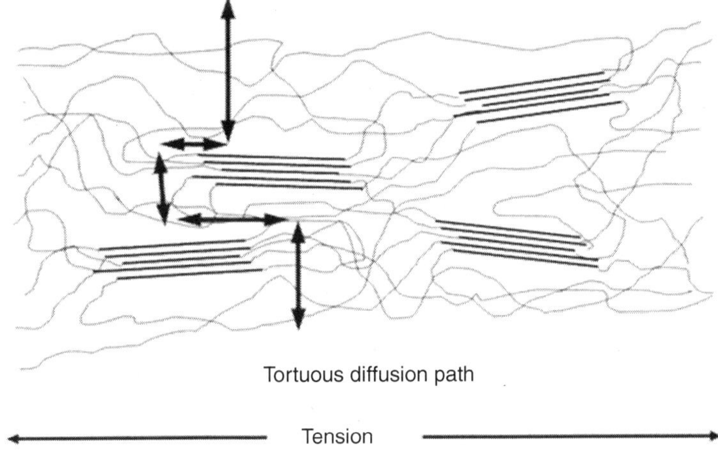

Figure 5.2 Oriented polymer film.

The orientation of the polymer takes the extruded thick polymer and as it is forward and sideways drawn the film stretches in these two directions but thins down in the third direction. The mechanical and optical performance of the films generally is improved by the orientation process. Properties such as increased tensile performance, reduced haze, higher gloss, higher tear and puncture resistance.

There are two main ways of converting extruded or cast film into bi-axially oriented film. Figure 5.3 shows a schematic of a sequential bi-axial orientation film line were the forwards draw and sideways or transverse draw is done separately. The forwards draw is achieved by having two sets of rolls that grip the film with the second set of rolls rotating faster than the first. The extruded film is thus stretched in the gap between the two sets of rolls. The first set of rolls is heated to make the film hot and easier to stretch. There is then a gap where the film is stretched followed by a second set of rolls that are divided into two parts. The first rolls are heated to help the polymer relax and reduce the tendency to shrink. The last of the rolls in this second set are cooled to freeze the properties into the film. Following this the film is then grabbed at the edges by the stenter clips. The film then enters the stenter or tenter oven and the film is re-heated and the clips pull the film sideways stretching it for the second time but in an

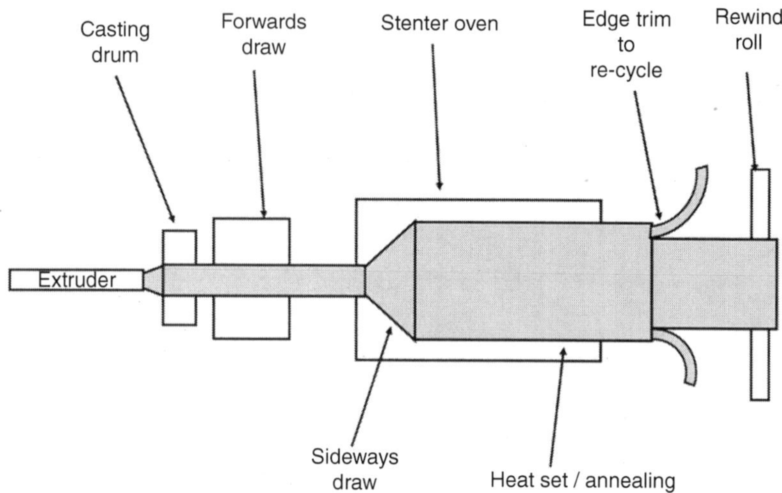

Figure 5.3 The bi-axial orientation of polymer film by a sequential drawing process.

orientation 90 degrees from the first stretch. The film again is held at a high temperature for a short time to allow for relaxation and then is quenched to set the properties into the film. As the edges of the film have been held by the clips the film is distorted at the edges and so this material is slit off and the edge trim is chopped up and re-enters the feed hopper into the extruder. The stretching in each direction can be by a factor of 3 – 5 or more depending on the polymer and precise properties required. It is also possible to have unbalanced stretching (i.e. 3x forward and 9x transverse) to provide specific properties. An example of uneven orientation is shrink sleeves where it is required that the film shrinks at a later date and conforms to a predetermined shape such as a bottle neck.

The film produced by this method has some residual stress left in the film and this residual stress is not uniform across the whole web but changes towards the edges. This can be improved by using a simultaneous bi-axial orientation process. There are stenters that can achieve this where the clips that grab the film edge are not only pulled sideways but also are pulled away from each other at the same time and so simultaneously stretch the film in both directions. It is mechanically more complex to pull the clips both sideways and forwards at the same time. This has been done mechanically by either a pantograph or spindle stenter or electromechanically by linear motors.

The alternative technology for simultaneous bi-axial orientation is the bubble process which is shown schematically in Fig. 5.4. In this case the extrusion, instead of a flat sheet as for the stenter process, is in the form of a tube. This tube has gas fed up the centre and the gas pressure inflates the tube into a bubble. The combination of the bubble inflation and draw off speed provide the bi-axial orientation. The extruded tube diameter and thickness and diameter and height of the bubble determine the amount of stretching that is put into the film. The height of the bubble has to be enough to allow the bubble to cool and solidify sufficiently that it can be collapsed down into a flat sheet. The collapsed bubble is in fact continuous until the edges are slit and the two sides can be separated and each wound up as a separate roll.

On either the stenter process or bubble process it is possible to produce co-extruded films. Co-extrusion offers the possibility of modifying the film properties by using different combinations of feed polymers to make multilayer films. The feed polymers can either be the same polymer type but with different amounts of additives or they can be different polymers with or without additives.

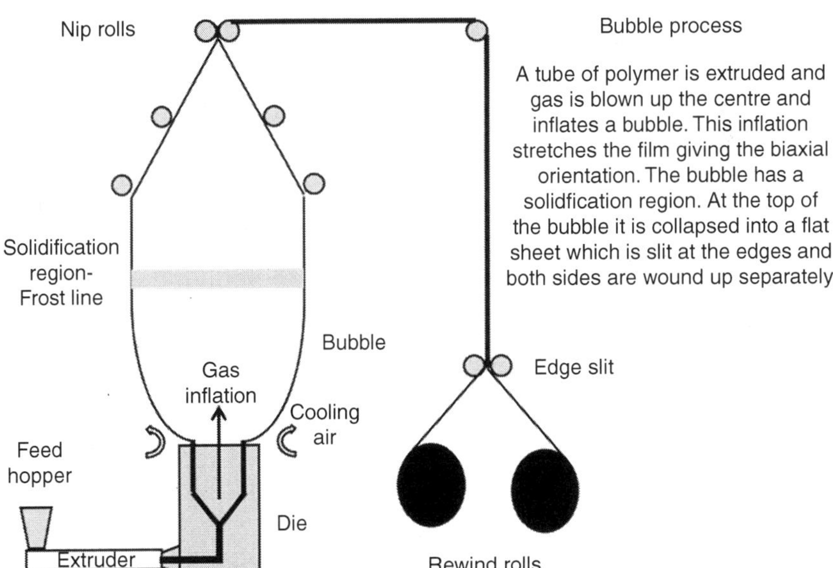

Figure 5.4 A schematic of Bubble Process for the bi-axial orientation of polymer film.

Polymer films can be very flat and if wound up can block and have serious winding problems. Blocking is where two polymer surfaces when brought into contact with each other cannot move or slip relative to each other. If pressure is applied to the films, such as would happen in a roll of film, any air separating the films may be removed and then the surfaces would stick to each other and it would then become impossible to separate the films from each other. To prevent this it is a common solution to add some filler to the polymer where some of the filler produces a surface roughness enabling the surfaces to be kept apart as well as reducing the coefficient of friction allowing the surfaces to move across each other more easily. The filler could reduce the optical properties of the film if used through the bulk of the film but if the filler were only used in a thin co-extruded layer with the bulk of the extruded polymer being without filler the optical properties would be closer to that of a completely unfilled polymer.

Co-extrusion of different polymers has been championed by manufacturers of the bubble process systems where it is possible to buy bubble lines that produce 5, 7, 9 or more layer co-extruded multilayer film. This can be from three or more extruders and so can be from multiple different polymers. The purpose of these multilayer multi-polymer films is to use the best properties of the different polymers to produce a single film with superior properties over any single polymer film.

5.1.1 Oligomers

The barrier performance of vacuum deposited coatings is dependant upon the quality of the substrate supplied. The polymerisation process is never perfect and there is always some proportion of unpolymerised monomer present in the film produced by the extrusion process (2–6). As the polymerisation process has become better understood the oligomer levels have reduced. Typically oligomer levels have been in the range 1%–3%. Established manufacturers, with their long experience, tend to produce material with lower oligomer levels than the newer low cost film suppliers. It is possible to purchase film with very low extractable oligomer levels in the region of <0.5% but these films take longer to produce and so are sold at a premium. In polymers such as polyester terephthalate (PET) there will also be a quantity of moisture also within the bulk of the film (7). The amount of moisture will depend on the polymerisation process and the speed of processing. It is possible to produce very dry films but it takes longer and this adds to the film cost (8).

The unpolymerised monomer can be very mobile and can migrate through the film and so will be present on the surface of the film (9–13). This material can be present as multiples of individual quantities of material referred to as mer units. These can be present as dimers, trimers or in larger groupings and all of the material is generically referred to as oligomer.

Figure 5.5 shows the surface of some PET that has been heated to 140°C and then has been photographed under a magnification of 2000x using scanning electron microscopy.

This oligomer material tends to be of a lower melting point than the polymer and is very mobile on the surface. Oligomer being mobile on the surface will move around and if there is debris on the surface can accumulate around debris or protruding filler particles. Hence often where there is a pinhole following the removal of some debris there may also be evidence of oligomer in the pinhole too. If a coating is vacuum deposited onto the polymer surface where there is oligomer present the adhesion will be poor. The coating may adhere to the oligomer but the oligomer has poor adhesion to the polymer surface and so adhesion failure will still

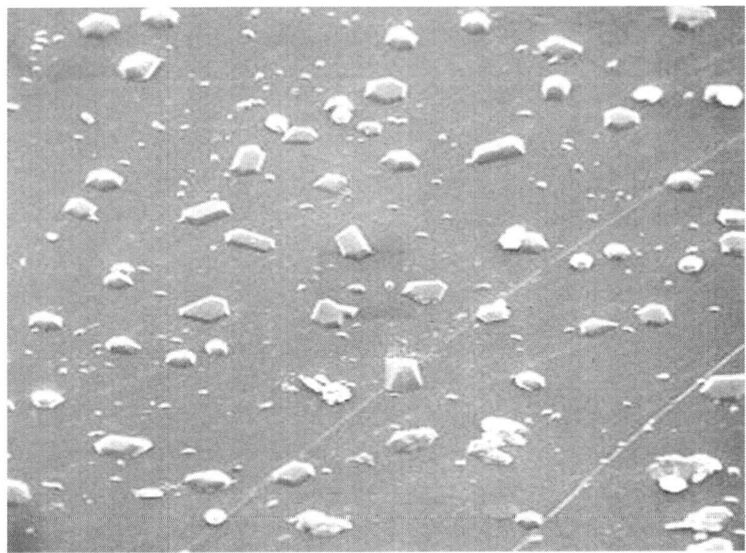

Figure 5.5 Oligomer deliberately grown large by prolonged heating. SEM 2000x magnification.

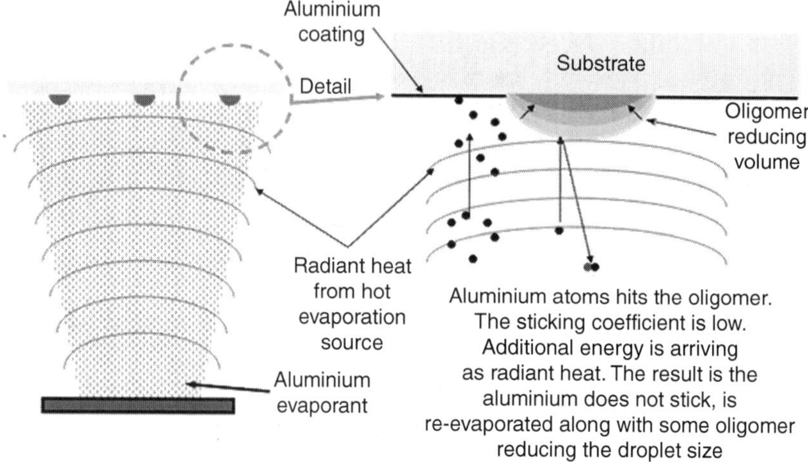

Figure 5.6 A schematic of how oligomer can result in a pinhole in the coating.

be a problem. Oligomers can also be a source of pinholes. When the film reaches the deposition zone the heat load from the deposition source can heat the oligomer to the point where the oligomer evaporates from the surface preventing deposition of the coating and so leaving an area of either reduced or no coating where the oligomer was present. Oligomers can sublime at temperatures well below the melting point enabling this process to occur at temperatures achieved by the polymer surface in the deposition zone. This process is shown schematically in Fig. 5.6. This process is exactly the same as is used in the in-vacuum pattern coating process where oil is printed and the oil prevents the deposition but is evaporated by the process leaving the film dry but uncoated where the oil has been present.

5.1.2 Additives

Some additives can cause the same problems as oligomers. Additives can be used for a variety of benefits starting as an aid to the polymer making process through to providing functional benefits in the final product. This number of additives has increased over the years and now includes the following:

Anti-oxidation
Nucleation promoters

Radical scavengers
UV absorbers

- Cross-linking agents
- Surface active agents
- Anti-misting
- Slip agents or lubricants
- Colour – pigments or dyes
- Whitening agents – optical brighteners
- Viscosity modifiers
- Gloss improvers
- Anti-static
- Anti-blocking fillers
- Voiding agents
- Dispersion stabilisers
- Flame retardants
- Foaming agents

Examples of where additives might be used would be the addition of a nucleation additive such as benzoate salts (sodium, aluminium, potassium) or sorbitols are used at levels of up to 0.5% in PP for the crystallisation process, which controls stiffness and clarity. PP is prone to oxidation and so phenolic or phosphite type anti-oxidants can be added in an amount between 0.01% to 0.5% to limit this oxidation. Also anti-static additives such as glycerol monoesters or ethoxylated secondary amines are used at concentrations between 0.1% and 1.0%. These additives are hydrophilic and also have limited solubility in the polymer and so they too migrate to the surface where they are most effective.

Of the additives used the one that stands out as being problematic to vacuum deposition is the slip agents. These additives are used to reduce the coefficient of friction (CoF) to make the film more easily handled. They are added to the bulk of the polymer and are designed to be able to migrate to the surface of the film. The speed that the slip agents migrate to the surface will depend on the time and temperature as well as the concentration in the polymer. Figure 5.7 shows a schematic of how the slip agent operates. Initially there is little or nothing present on the surface. With time and temperature some slip agent will start to appear on the surface and the surface energy of the polymer surface will decline. Eventually enough slip agent will migrate to the surface that a monolayer will completely cover the polymer surface and at this point the surface energy and CoF will reach a minimum. Using low density polyethylene as an example the CoF might be as high as 0.7 but with the addition of 0.1% slip agent this can be reduced to a CoF of 0.2 where a CoF of 0.16–0.2 is regarded as the range that provides good film handling properties. This reduction in surface energy will reduce the wetting of any depositing coating which will make the barrier performance worse as well as reducing the adhesion. Reduced adhesion is often also seen as a higher level of pinholes and that also makes the barrier performance worse.

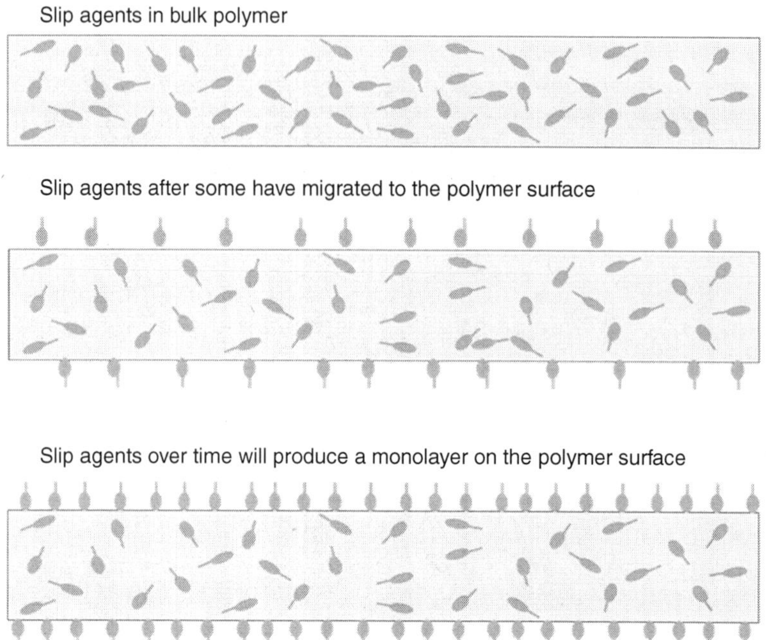

Figure 5.7 The migration of slip agents with time from the polymer bulk to the surface.

Some slip agents are produced by the amidisation of long chain fatty acids with the most commonly used of these being Steramides, Erucamides & Oleamides.

Stearamide is prepared by amidation of stearic acid (C18:0).	Mpt. 98–104°C
Oleamide is prepared by amidation of oleic acid (C18:1).	Mpt. 66–76°C
Erucamide is prepared by amidation of erucic acid (C22:1).	Mpt 79–85°C

Erucamides are longer chain, more heat stable and more oxidation resistant than Oleamides and with a lower vapour pressure create fewer volatiles during high temperature processing. The Oleamides migrate through to the surface more quickly and are sometimes referred to as 'fast blooming'. The aim is to provide sufficient slip additive so that with a suitable time for migration no more than a monolayer will have accumulated at the surface. More than a monolayer does not make any further improvements on the slip performance.

Waxes can also be used as slip agents. Waxes are similar to oils except they are solid at ambient temperature and generally have a melting point in excess of 40°C. A very interesting evaluation study with microcrystalline waxes clearly demonstrated that the harder the wax, the better the slip properties.

Fillers are used to roughen the polymer surface to reduce the contact area and by doing so reduce the Coefficient of Friction. In increasing the surface roughness they also increase the haze and reduce the specular reflectivity. Typically the haze increases somewhere between 0.4 and 1.0% per 1000 ppm of silica filler used. There is a balance between the amount of filler used to make handling easier and the optical properties. The reduction in CoF using fillers will take the CoF down to around 0.3–0.4 which is not enough to make handling easy enough and so it is typical to find slip agents added as well as fillers.

Fillers can vary in size, shape and material. Below are a few that are commonly used.

Synthetic silica – high surface area, hydroxylated and microporous surface. A good match of refractive index for PE & PP and used to make highly transparent films.

Natural silica – a different mixture of shapes and sizes and often contains impurities that the synthetic silica does not. Cheaper than synthetic silica.

Talc – Magnesium hydrosilicate, a lamella type of soft rock that has a refractive index that is also a good match to PE & PP.

Limestone – A low cost filler. Calcium carbonate and sometimes a mixture with magnesium carbonate these are used in lower quality film applications.

The fillers reduce the contact area and the slip agents add a lubricant and in combination reduce the CoF sufficiently to make handling easy. Unfortunately this makes the deposition of vacuum coatings more difficult. To get sufficient adhesion the polymer surface has to be treated to try to either remove the low energy mobile molecules from the surface or alternatively crosslink the mobile molecules into the polymer bulk. If the adhesion is not good enough then what can happen is shown schematically in Fig. 5.8 below.

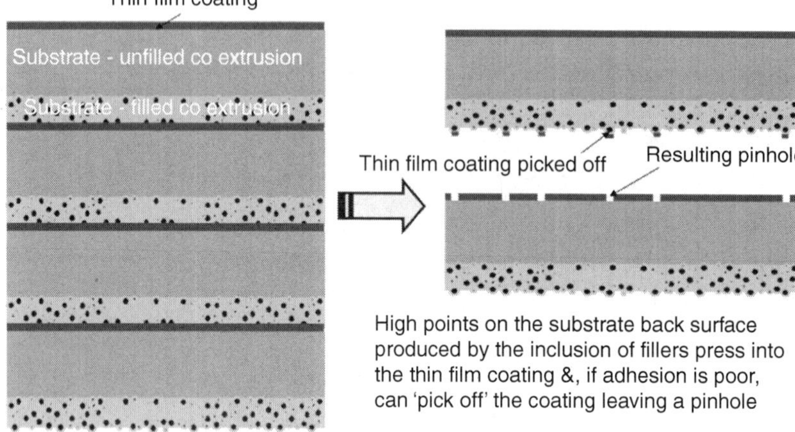

Figure 5.8 The use of the co-extrusion of a thin filled layer to reduce the Coefficient of Friction (CoF) by roughening one surface. The high spots can lead to pick-off pinholes if adhesion is poor.

The filler from the back surface of the film that gets wound onto the coated surface is pressed into the coated surface. It is possible that the high surface energy of the deposited coating and the pressure enables the coating to stick to the high spot caused by the filler on the back surface. When the film is separated if the adhesion is higher to the filler high spot than to the original substrate surface then the coating will be picked off the substrate and will stay on the filler high spot on the back surface of the layer above. The filler near the surface may be covered by a thin polymer skin or may also break through the polymer surface. Those that break through lead to a disruption in the surface smoothness and can lead to coatings being uneven and stress cracking later when the film is flexed as shown later in Fig. 5.11. This type of defect worsens the barrier performance and so filler type and quantity is important (14–16).

The slip agents may get removed from the substrate surface before vacuum coating but invariably they are still present on the back surface of the substrate. This can lead to two problems one is that if the coating surface is cleaned before vacuum coating but too much time is allowed between cleaning and coating it is possible that the slip agents re-appear before the coating takes place. The time between cleaning and coating is also dependent upon the temperature the roll of film sees over this storage period. The higher the temperature the shorter the time in which the surface will become re-contaminated. This will also be dependant upon the polymer type and crystallinity as the higher the crystallinity the slower the migration because of the higher density of the crystalline polymer as shown in Fig. 7.2 and the more tortuous path. The slip agents can also be chosen to be either shorter chain lengths and so quicker to migrate, the fast blooming slip agents, or they can be chosen to have a longer chain length and be slower to migrate.

Once the slip agents have been cleaned off the front surface and the vacuum deposited coating added the problem is not ended. The freshly deposited coating has a very high surface energy and this will be wound up and so in close proximity to the back surface of the same film which will still have the slip agent over the surface. This slip agent on the back surface is highly mobile and as nature always tries to bring surfaces to equilibrium the slip agent will migrate across from the back surface to the contacting front surface and spread out from these contact points lowering the surface energy of the freshly coated surface. Eventually the front surface vacuum deposited coating surface energy will decline to match the back surface energy level. Again the time for this to occur will depend on the chain length and hence mobility of the slip agent as well as the storage temperature.

5.1.3 Contamination

Contamination of the polymer film surfaces starts immediately the polymer is extruded. Unpolymerised monomer can be volatile and any of this that is released from the polymer surface can condense on nearby cool surfaces. This condensate will form a white powder and over time this powder will accumulate and start to be released where it can either join the atmosphere or fall directly onto the polymer web. This occurs also in the stenter oven and in some places this powder is referred to as snow because periodically it can be seen falling onto the web. Polymer films will also produce a triboelectric charge as the film approaches and passes over rollers (17–20). Any two dissimilar materials when brought into contact will use the movement of electrons to try to bring both surfaces to equilibrium. The pair of surfaces will have one positively charged and the other negatively charged. When these two surfaces are separated the surfaces will retain the charge present. If the materials are conducting a current will flow and the surface be discharged but if the materials are non-conducting the charge will remain. A polymer can act as a capacitor which will result

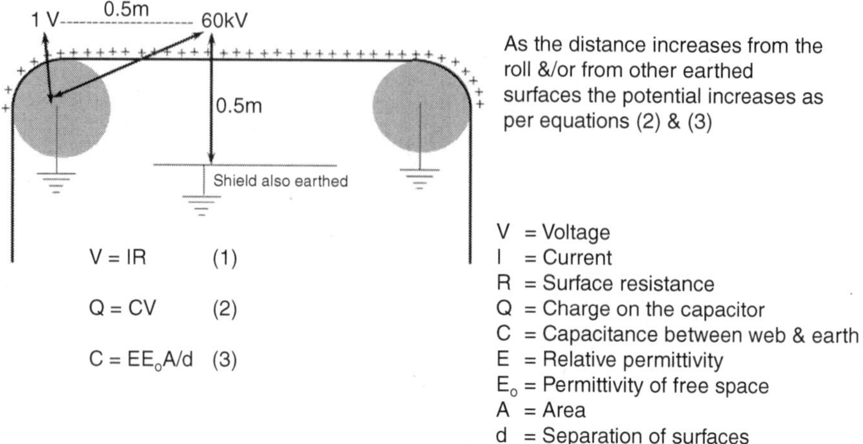

Figure 5.9 The triboelectric charging of polymer films passing over metal rolls.

in the surface charge increasing as the polymer increases in distance away from the rolls. This is shown schematically in Fig. 5.9.

This triboelectric charge can also attract dust/debris out of the atmosphere effectively acting as an electrophoretic collector (21). There is always debris present in the atmosphere from a huge variety of sources from pollen, mechanical debris from any moving parts, flaking paint, human detritus such as shed skin or hair, dust from cardboard cores or packaging material besides the polymer debris (22). Further down the production process there is the slitting process where mechanical blades cut the polymer. Unfortunately this process generates dust of which the amount will depend on how sharp and well set the blades are.

This collection of debris by the film is cumulative. The more the film is handled the higher the level of contamination. It was observed by the author that the dust levels were measured at approximately double the previous levels following slitting. This was despite having a vacuum extract to try to minimise the dust escaping into the atmosphere from the slitting process.

Even without the attracting force of the static charge on the polymer surface debris will still fall onto the surface of the polymer as particles naturally fall under gravity and will land on any horizontal surface such as the polymer film. The larger the particulate size the faster the particles will fall down and collect onto the film.

Table 5.2 Dust settling rates.

Settling Rates for airborne particulates		
Diameter of particles	**Velocity of settling**	
microns	**feet per minute**	**mm per second**
1	0.007	0.0004
5	0.2	0.0109
10	0.59	0.0323
60	21.3	1.1644

Table 5.2 gives an indication of the settling rates for particles of different diameters.

The debris is a major source of loss of barrier. The debris passes through the deposition zone and is coated along with the polymer film. The film as it is wound up or passes over front surface rolls may, or may not, move some of the debris. Where the debris is moved it leaves behind a small area of uncoated film which is known as a pinhole (23). If the debris is rolled away from the original position the shape of the pinhole will approximate the same shape as the debris that was moved with the coating thinning from full thickness coating down to no coating under the particle. This thinning is due to the shadowing effect of the debris as it enters, passes through and exits the deposition zone. If the debris is moved in a sliding motion away from the original site the pinhole will also have a scratch away from the pinhole which may end at the debris if it is still present. This is shown schematically in Fig. 5.10.

Debris not only is a direct problem to the deposition of good barrier coatings (24) but it is also a potential source of web wrinkles (25). It has been shown that a particle as small as 5 microns on the back surface of the film will keep the polymer from making good contact to the cooled deposition drum (26). This will locally reduce the heat transfer coefficient and this can cause the film to expand faster than the surrounding film and it can initiate a wrinkle. Thus it is important when cleaning the film to make sure both sides are cleaned and not just the front surface.

The polymer film manufacturing process is not an easy process to try to clean up. Film lines are run semi-continuously which makes

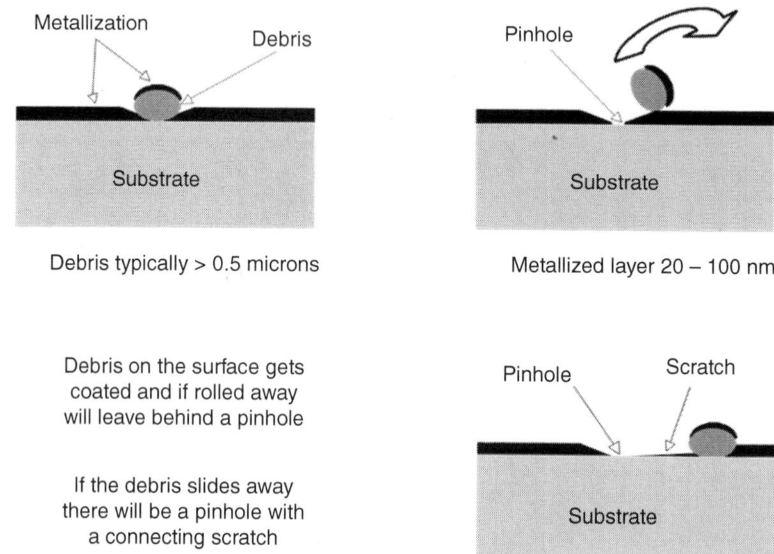

Figure 5.10 How debris contamination on the polymer surface results in pinholes.

it difficult to keep the powder levels constant and to a minimum. Particularly in the stenter ovens where hot air is circulated the powder gets stirred up and the volume of air is so great it is impossible to filter out the powder. If the filters are fine they become clogged up too rapidly and block the air flow and if coarser filters are used they do not catch enough of the powder to make much improvement. If the air were to be used only once the costs would go up significantly for the cost of heating new air each time. Reducing the airborne debris that falls onto the exposed film surface is possible by using a combination of electrostatic neutralisation of the polymer film surface and using clean air positive pressure hoods. These localised hoods limit the quantity of airborne debris that can fall onto the web by covering these exposed areas with a hood filled with clean filtered air.

It is unrealistic to expect film suppliers to produce perfectly clean film at the same cost as existing product. However what can be expected is that as the suppliers know that the film is contaminated they will be increasingly expected to clean as much of the debris off as they can before winding the polymer roll up. Some already do this as part of their product differentiation.

5.1.4 Surface Quality

In the quest to improve the vacuum deposited barrier coatings from food packaging barrier into ultra barrier materials it has been shown that the surface quality of the polymer film is important. Surface roughness has been directly related to coating defects and barrier performance (27–29). If the surface is rough the peaks and valleys can become stress raisers and result in cracking of the vacuum deposited coating. This surface roughness can be as a direct result of any fillers used in the polymer composition. It is not just the peaks formed by the fillers that can cause cracking in coatings but also dents in the surface. Dents can be caused by the indentation of fillers from the back surface of the contacting film in the polymer roll. The higher the roll tension and the harder the wind the greater will be the number of this type of surface defect. A schematic of where cracking can occur in the vacuum deposited coatings is shown in Fig. 5.11. Anywhere there is a change in the surface from being flat there is the potential for an increase in the coating stress particularly when the coated film is flexed. One of the tests for barrier coatings is to flex the coated material over a series of mandrels of decreasing diameter and test the barrier performance after each one. There will often be a mandrel diameter where the stresses cause a sudden increase in the amount of cracking which is reflecting in a worsening of the barrier performance. The smaller the mandrel size before this cracking happens indicates the higher the coating adhesion.

The surface of the polymer film has many components. There is the surface roughness that relates to the polymer composition.

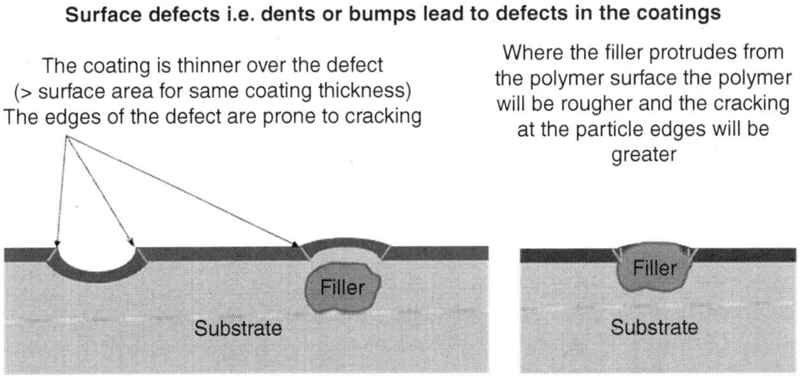

Figure 5.11 Surface defects resulting in cracking following film flexing.

Chemically it is often expected that if the polymer is polypropylene then the surface chemistry will be that of polypropylene but this is rarely if ever true. There may be additives that migrate to the surface as well as the oligomer that can migrate to the surface. With some polyester films there may be coatings applied during the film manufacturing sometimes at the interdraw position between the forwards and sideways draw on the film line. These coatings are very thin but are chemically very different to polyester and are used to improve the handling of the film or as adhesion promoters for specific coating materials. In addition there is then the contamination from the debris collected from the atmosphere and made worse by the charged film surface. It is important to note that changing supplier will most likely give you a change in the surface that you will have to work with. The polymer type will be the same but their manufacturing process may have small differences that will give differences in oligomer and moisture content and any additives may well be chemically different and be present in different proportions. Similarly any coatings are likely to be proprietary and chemically different. So changing film supplier can mean that the surface quality, contamination levels, surface chemistry and hence the coating adhesion may well be affected too.

Managing the substrate surface is one of the key aspects of producing the best barrier performance possible. Often for packaging applications there is little choice of what type of film is to be used. The polymer type and thickness may be defined by the customer. The treatment of the polymer surface then becomes the only method of differentiating vacuum deposition coaters. The surface can be cleaned to remove contamination, coated to redefine the surface and plasma treated to chemically modify the surface. The aim is to achieve the best possible performance at the least cost.

5.2 Substrate Cleaning

Although the ideal is not to contaminate the polymer surface in the first place we have seen that this is not possible and so films have to be cleaned. There are many possible cleaning methods but a number of them can put damage into the film surface which will also be a cause of worsening the barrier properties of any vacuum deposited coating.

The following are some of the possible cleaning techniques;

Cloth wipe
Static brush
Rotating brush
Air blower / knife
Vacuum cleaner
Combined ultrasonic air jet & vacuum
Transfer tack rolls
Coating
Chemical bath
Carbon Dioxide 'snow jet'

The physical contact methods such as the cloth wipe, static brush and rotating brush are methods used in the glass coating industry where the substrate is hard and resistant to scratching. Polymer films are very soft by comparison and the debris on the surface is often much harder than the polymer and so is abrasive. This abrasion puts many scratches into the polymer surface which is unacceptable.

The air blower and vacuum cleaner techniques have the problem that there is always a boundary layer of air that travels with the film on each surface. This boundary layer relates to the winding speed of the film but is usually a few tens of microns. The vacuum cleaner or air blower have difficulty in penetrating this boundary layer and so most of the debris remains in place with only the very largest debris being removed.

The combined ultrasonic air jet and vacuum cleaning process overcomes this boundary layer effect because of the ultrasonic pulsing of the air jet. The ultrasonic pulsing vibrates the polymer web making it flutter and as the film changes direction the debris continues in the same direction and so is lifted off the film surface and so is more easily able to be vacuumed away. This process has been used to remove all debris down to a diameter of 0.3 microns. This process is quite noisy with the large quantity of blown air and substantial vacuum extract.

The chemical baths tend to be expensive and the cost is growing as the disposal costs of liquids are increasing. The film has to be wound through the cleaning bath which may contain either solvents or detergents to aid the release of the debris from the surface. There need to be additional baths that then wash off any of the solvents or detergents to return the surface to a clean state. The film then needs to be dried otherwise there may be stains left on the surface or an increased moisture content which will slow down the vacuum processing time. Where this process has been shown to have some advantage is where it is possible to chemically etch the surface to increase the coating adhesion. Usually the costs are prohibitive and alternative cleaning processes easier and more cost effective.

The carbon dioxide or snow jet process has been used on small polymer components such as spectacle lenses. Although this process could be used on larger substrates the costs again make it less attractive than some of the other cleaning processes.

The tacky roll cleaning process is a contact cleaning process that does not damage the substrate. It has been used in the display industry prior to lamination processes successfully where cleaning debris from mating surfaces prevented Newton rings being visible in the display screen. This cleaning method is also the one cleaning process that can be used inside a vacuum system. The other cleaning methods need to be done at atmospheric pressure and so either need to be included on one of the earlier processes or needs a dedicated winder just for cleaning the film. Some manufacturers are now including a tacky roll cleaner just before the re-wind roll so that they clean as much of the debris off the web before it is re-wound. It is important to make sure the surface is not re-contaminated following cleaning. If the film passes over more rollers before it reaches the re-wind it will be building up more surface charge and attracting more debris to the surface following cleaning. Hence it is important to either have the cleaning immediately before the re-wind or to use clean filtered air positive pressure hoods to protect the film from re-contamination. The way the tacky roll works (30,31) is shown schematically in Fig. 5.12.

The roll contacting the web is tacky enough to remove the debris down to approximately 0.3 microns. The opposite side of this tacky roll contacts a higher adhesive roll that removes all of this debris. This high tack or adhesive roll accumulates the debris and so over time the ability to keep removing the debris can decline and so periodically the adhesive surface is refreshed. This adhesive roll is built up as a series of layers so that to refresh the roll it is simply a matter of peeling off the exhausted layer. The benefit of this type of cleaning is shown below in Fig. 5.13. This technology is still being developed with one manufacturer now offering a tacky roll process that is capable of removing debris of a much smaller size than by any other technique (32).

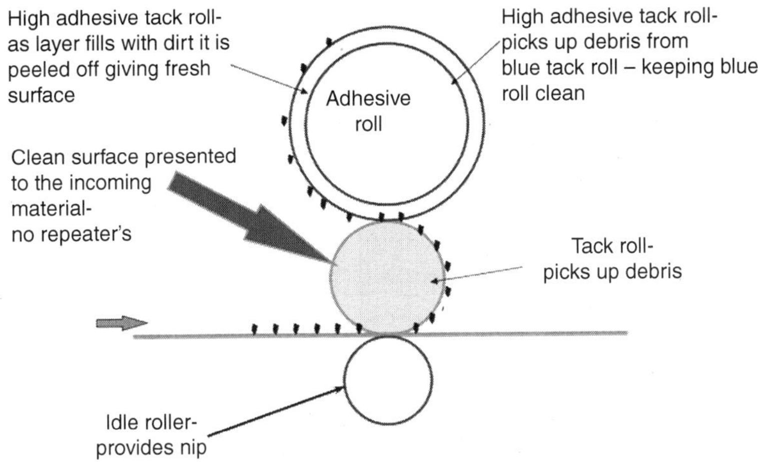

Figure 5.12 A schematic of single side tacky roll cleaning of a flexible film.

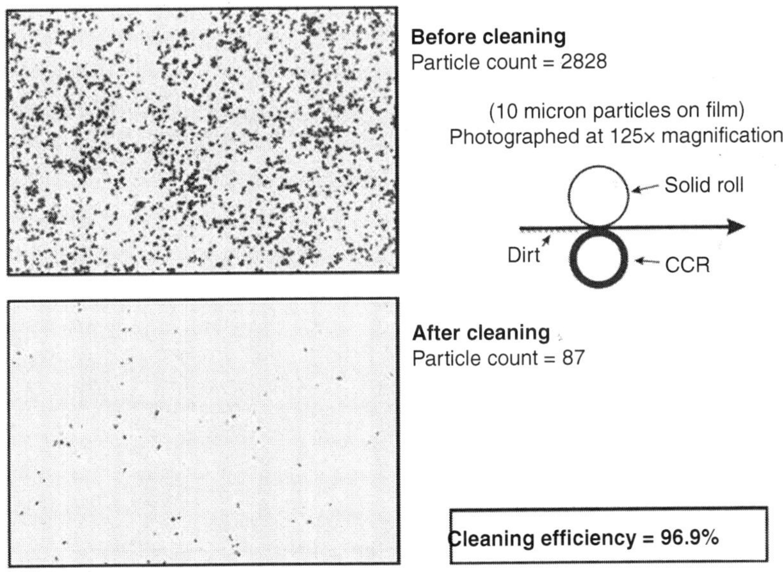

Figure 5.13 Images courtesy of Polymag Tek Inc. demonstrating the benefit of tacky roll cleaning (33).

There are problems of using this technology in vacuum systems. It is important to use the right type of tacky rolls that have vacuum compatible materials. One supplier worked with NASA to provide the technology to be used in space and so have vacuum compatible materials. If non-vacuum compatible materials are used the tacky rolls can dry out and become less effective. The other difficulty in using the process in vacuum systems is the refreshing of the high tack or adhesive roll. Normally an operator intervenes and tears off the exhausted top layer but this is not possible if the system is in a vacuum. Currently those using the process in vacuum either have a limited roll length and so do not have to refresh the high tack roll or they use a much higher surface area than conventionally to extend the lifetime of the high tack roll.

Cleaning should be done on both surfaces rather than one surface only as when only one surface is cleaned debris from the back surface may be transferred to the clean surface once the film is wound up into a roll.

Cleaning particles becomes increasingly difficult as the particle size gets smaller. The reason for this is that the method of the particle being held onto the surface changes.

As can be seen in Table 5.3 that gravity plays a very large part in the adhesion of large particles above 10 microns to the surface whereas once the size is down below 1 micron the adhesion is primarily by the Van der Waals forces. These forces are very strong and it becomes increasingly difficult to pull each particle off the surface (34).

5.3 Substrate Plasma Treatments

Substrates can be pre-treated before they are vacuum coated. The reason for pre-treating the polymer surfaces is to change the surface properties with a view to increasing the adhesion and performance of the vacuum deposited coating.

Table 5.3 The forces holding debris onto surfaces.

Debris ave. diameter	**1 micron**		**10 micron**		**100 micron**	
Force type	**Force - millidynes**	%	**Force - millidynes**	%	**Force - millidynes**	%
Van der Waals	0.4	99	4	97	40	29
Electrostatic	0.005	1	0.05	1	0.5	
Gravity	0.0001		0.1	2	100	71
Totals	0.4		4.1		140	

Plasma treatments can be of a variety of different types such as flame, corona, atmospheric plasma or vacuum plasma (35–47). All these plasma types contain ions and electrons in different proportions and it is the energetic species that can modify the polymer surfaces. The heavy ions may break bonds and may even sputter the surface and then depending on the gases present there may be chemical modification to the surface (48–61). In most cases oxygen is used as part of the gas composition. The oxygen has two functions, one is to produce volatiles out of hydrocarbon surface contamination that may then be pumped away. Secondly the oxygen may be bound to any free bonds produced by the scission of the polymer chains. This bonded oxygen will then be available for the depositing coating to bond to, making the adhesion stronger. Each of the plasma types has slightly different attributes that make the treatment slightly different. The lifetime of the changes may also vary depending on the amount of treatment as well as the speed of return of any migratory additives or oligomer. The polymer surface is often different to the bulk polymer and may be a mixture of crystalline and amorphous regions each of which may be affected differently. Similarly the time for the surface effects to decay can be different for different polymers. It is believed that polymer chains can slowly move such as by rotation to reduce the surface energy with the speed of movement being polymer dependent. Amorphous polyester terephthalate is more mobile than polypropylene which is thought to contribute to the difference in treatment lifetime even where additives are not being used or the oligomer content is low.

It is important to note that most of these pre-treatments if used well can improve the surface properties but, if not optimised, it is also possible to degrade the polymer surface either retaining the original poor adhesion or even making the adhesion worse (62,63). Where the vacuum deposition system has a vacuum plasma treatment process available before the deposition process it is usual for this to be used. This may be used on substrate film that has already had a previous atmospheric pre-treatment and this can make it difficult to truly optimise the treatment unless the history of the treatment and storage conditions and time since the treatment are known. In some cases where the storage time is minimised the vacuum plasma treatment may be too much and reduce the adhesion because of over treatment of the surface by the combination of both treatments. However if the film has been stored for some time and the oligomer or additives have re-populated the surface then the vacuum plasma treatment may be optimised and the adhesion at a maximum. This is why a full knowledge of the substrate history can become important to enable reproducible production.

Flame treatment uses a hydrocarbon gas and air mixture and depending on the gas proportions and positioning of the polymer in the flame the polymer may see either an oxidising or reducing flame. Control of the gas type, gas ratio, gas flow, residence time and position of the film in the flame will all have an effect on the final chemistry of the flame treatment (63–65). Flame treatment is shown schematically in Fig. 5.14. The flame treatment can oxidise the surface and the treatment will occur down to the depth of a few nanometres and the treatment can be quite stable and in some cases less likely to decrease with time compared to the corona treatment.

The flame is hot and will damage the film very rapidly and so it is important that the film is kept moving whilst the source is ignited and the flame present. The film should also be treated with the film passing over a cooled roll. The flame source is often a single source with the gas diffused through a series of corrugated plates so that there will be an array of flame that gives a uniform treatment across the film width. The flame stability has been improved with more precise control of the gas flows and the source will have an auto-ignition unit to make sure that the flame does not extinguish.

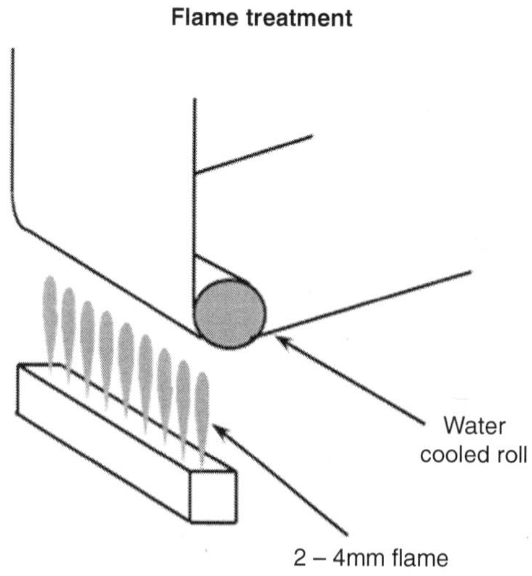

Figure 5.14 Flame treatment.

Flame has the additional advantage over corona in that it cannot create a plasma behind the film nor does it arc and so does not produce either a back surface effect or pinholes in the film.

Corona treatment, as shown schematically in Fig. 5.15, uses a high voltage and high frequency to break down the air and create a plasma across an air gap that acts as a capacitor. The voltage is raised to a point where the air breaks down and ions and electrons are produced and a current can flow across the gap creating a self-sustaining plasma. The high frequency is used to help smooth out variations and give a continuous and uniform surface treatment of the film as it passes through the plasma. The metal roll, used as one half of the circuit, usually has a dielectric sleeve to increase the voltage drop across the plasma to reduce the arcing to the metal roll (66). The corona plasma generates ozone from the air and the ozone has quite an aggressive scrubbing action which can be good for removing hydrocarbon contamination but may roughen the surface if the treatment power is too high or residence time too long. This ozone requires that the corona treatment units have a suitable extraction

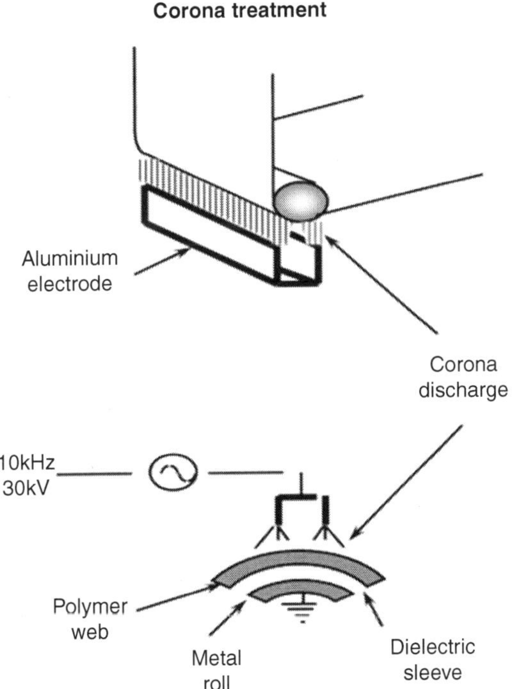

Figure 5.15 Corona treatment.

system for safety as ozone is a known carcinogen. Although the corona can raise the surface energy of the film surface it can leave behind a surface that is water soluble. The corona plasma can be subject to arcing (67) due to instabilities and this can be made worse by the surface quality of the electrode. These arcs may be severe enough to burn holes through the polymer web. As there is always some entrained air on the back side of the film it is also possible to cause some treatment to the back surface of the film too. This is often regarded as a problem rather than an advantage. A common problem with corona treatment is that it may not be optimised. Some corona treaters do not have a continuous power control but have step changes in power. The corona treatment can be adversely affected by changes in ambient humidity and so seasonal changes can mean that a corona treatment set up in the summer may not work as well in winter or through the monsoon season.

Of the atmospheric pre-treatments available corona is the one that is most widely used. This may well be about to change as atmospheric plasma treatment systems are becoming more widely available. The atmospheric plasma systems have been developed for a variety of reasons of which one reason was to eliminate the arcs that are produced in the corona treatment systems. This is done by controlling the frequency of the power supply and the gas composition and in doing so operating the system at a lower voltage than for the corona treatment. Many of the systems use helium as the main gas. The higher power supply frequency and the increased mobility of using helium means that as soon as an arc is initiated the switch in polarity and higher gas and electron mobility quenches the arc before it can become established. This makes the atmospheric plasma treatment attractive as the polymer substrates are not punctured by any arcs as can happen with corona treatment, as shown schematically in Fig. 5.16. With careful management of the gas flows any back surface treatment can also be eliminated. The advantage of this process is that other gases can be used in the mixture and the chemistry of the surface modified more widely and more easily than by the other methods. This includes being able to directly deposit coatings by a combination of decomposition and recombination or oxidation processes (68–70). It is possible to directly deposit some barrier coatings by this process without the need of any subsequent vacuum deposition process. Currently these coatings are thin and need either multiple process plasmas or the film needs to be passed multiple times past a single head.

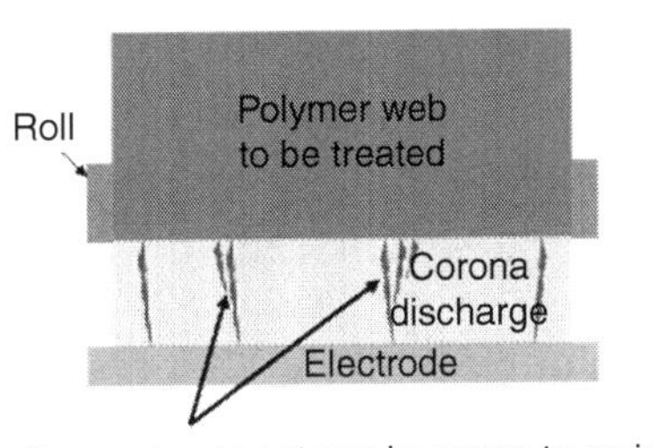

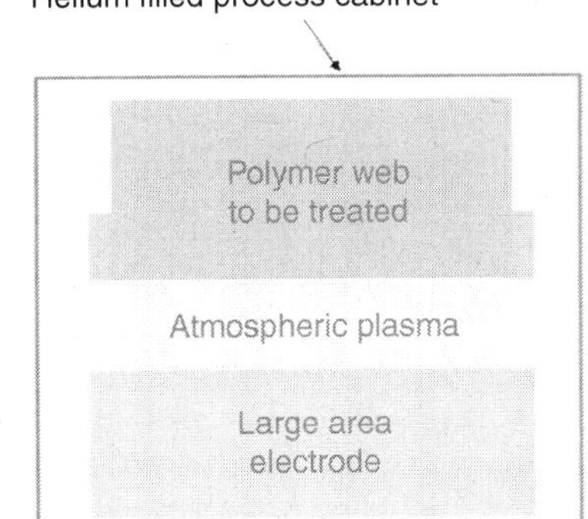

Figure 5.16 A schematic comparing corona to atmospheric plasma treatment.

The cost of helium and either multiple processes or passes does not make this process economic as yet but development is being done to make atmospheric plasma processing competitive to vacuum deposition processes. The work is aimed at either recirculating the high cost helium or to control the arcing using the power supply without the need for helium. These improvements will increase the use of atmospheric plasma treatment and coating.

The atmospheric plasma treatments age just like the other treatments and so for all of these processes the time and storage conditions between treatment and coating are important parameters.

The vacuum plasma treatment has the benefit of being at a lower pressure and hence is regarded as a low temperature plasma treatment. The voltage to sustain the plasma is lower especially if magnetic enhancement of the plasma is used. The magnets cause the electrons to travel in a spiral path and so the chance of undergoing an ionising collision is enhanced. This increased ionisation efficiency produces many more electrons and this enables the plasma to carry more current at a lower voltage. This has enabled more intense plasmas to be used enabling more compact plasma treatment systems to be designed and even though the winding speeds have increased and hence the residence times reduced there is still sufficient power to still fully treat the film easily. In fact the magnetically enhanced plasma treaters have to be used carefully when winding film slowly as there is still sufficient energy to thermally damage the substrate. Vacuum plasma treatments have been shown to smooth the rough polymer surfaces with the plasma removing some of the peaks but in other conditions too much plasma treatment causes surface roughening. The roughening is more common where very smooth planarised films are used such as supplied for ultra barrier applications.

Most vacuum plasmas use a combination of argon and oxygen to treat the film surface although other combinations have been tried (71,72). The plasma generates many active species including vacuum ultraviolet light which is very effective at producing photocatalytic reactions. Argon alone cannot clean hydrocarbons from the polymer surfaces. The heavy argon can bombard the surface and cause chain scission or it can help promote crosslinking of hydrocarbons including bonding them into the polymer chains. The oxygen is less effective at this type of bombardment, being lighter, but the oxygen can bond to the polymer chain and also can bond to the aluminium. The oxygen can also react with hydrocarbons on the surface which are volatile and can be evolved and pumped away. Depending on the treatment the surface may either be carburised or be oxidised.

The aim is to clean the surface and also increase the oxidation at the surface which both increases the surface energy (Tables 5.4 & 5.5) and hence wetting of the depositing coating and increasing the bonding and adhesion (73). As with the other processes it is possible to under treat or over treat the surface and so it is worth spending time to optimise the plasma treatment (74–76). The use of surface energy as a measure (see Chapter 3) of the treatment is used for all of the pre-treatments and can be used, with care, to help optimise the process. The polymer surface energy will increase with treatment power or time up to a maximum and then the surface energy will stay at that plateau level with any further increase of power or time. If we do a plot of adhesion against the same increasing time or power we will see that the adhesion also increases initially and reaches a maximum around the same time the surface energy reaches a maximum. However, if we continue increasing power or time the adhesion does not remain at the plateau but instead immediately starts to fall. The reason for this is that the plasma causes chain scission and also adds oxygen and there is a balance between the two processes. Once the maximum adhesion is reached the chain scission begins to create too many short chain fragments and starts to form a weak boundary layer at the surface. Further plasma treatment just increases the

Table 5.4 The relationship between surface energy and chemistry.

	Surface Tension Dyne/cm (N/mm)	**Number & type of significant atoms**	
CF_3	5	FFF	The table on the left is a list of various polymer fragments with the relevant surface tension measurements. The influential elements & quantity is shown in the right-hand column.
CF_2	18	FF	
CF_2–CFH	22		
CH_3	24	HHH	As would be expected fluorine has the lowest energy (cf. PTFE).
CF_2–CH_2	25		
CH_2	31	HH	
CHOH–CH_2	35	O	Oxygen raises the surface energy & it is this that is commonly used in plasma treatments to improve the adhesion of coatings.
CClH–CH_2	38	Cl	
CCl_2–CH_2	40	ClCl	
CCl_2	43		
CO_2–CH_2	43	OO	

proportion of short polymer chain fragments and so the adhesion continues to decrease as shown schematically in Fig. 5.17.

Plasma treatment is not the same for all polymers or even the same polymer but different suppliers. Because the treatment is a

Table 5.5 The changing composition of OPP and the corresponding surface energy.

Surface energy of OPP for different surface compositions		
Surface energy (dynes/cm)	**XPS Atom% C**	**XPS Atom% O**
>55	85	15
46 – 50	87	13
38 – 42	91	9
37 – 40	92	8
40 – 44	91	9

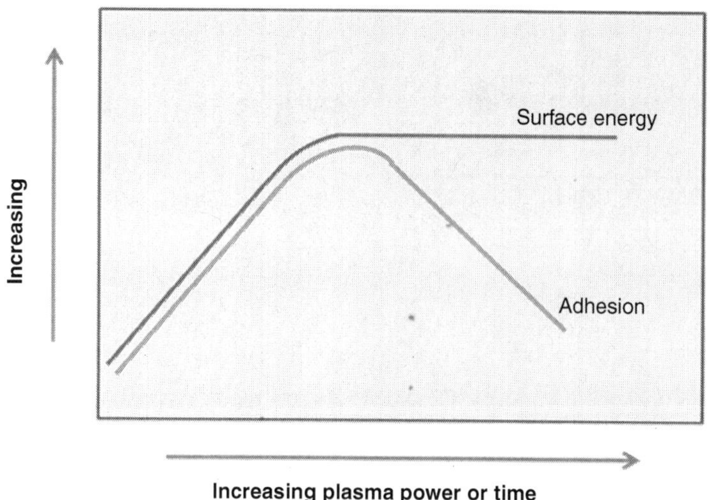

Figure 5.17 A schematic of the change in surface energy and adhesion with plasma treatment.

combination of bombardment that can sputter the surface or cause damage such as chain scission and then chemistry modification of the surface such as by increasing the oxidation the results of the plasma treatment will only be the same if the polymer and chemistry of the surface is identical. Changing supplier will often mean the surface chemistry changes because of proprietary additives or coatings and so the effectiveness of the plasma treatment can be expected to reflect these differences in a difference in the surface energy or adhesion for the same plasma treatment. Similarly some films that may already have had a corona treatment are sometimes given an additional vacuum plasma treatment. As the surface may degrade after corona treatment, as low molecular weight material either migrates to the surface or is transferred over from the back surface in the wound up roll, the material supplied to the vacuum system has to be regarded as a variable unless the time and storage conditions between pre-treatment and vacuum processing can be made a constant.

As with the other plasma treatments, the effects of vacuum plasma treatment may be short lived too (77). However as this treatment is usually done in the same system and as part of the same process, immediately before the deposition, it is less of a factor than for the other treatments.

It is widely believed that plasma treatment can remove debris from the surface because the plasma neutralises the surface charge and once this charge is removed the debris will fall away under the effect of gravity. Table 5.3 shows that the electrostatic charge has only a small effect on the adhesion of particles to the surface and that it is really dominated by Van der Waals force for particles of less than 1 micron and gravity for particle sizes larger than this. This means that relatively few particles will be released simply by removing the electrostatic charge. Figure 5.18 shows how debris can be attached to the surface by gravity, electrostatic charge, glued in place by mois-

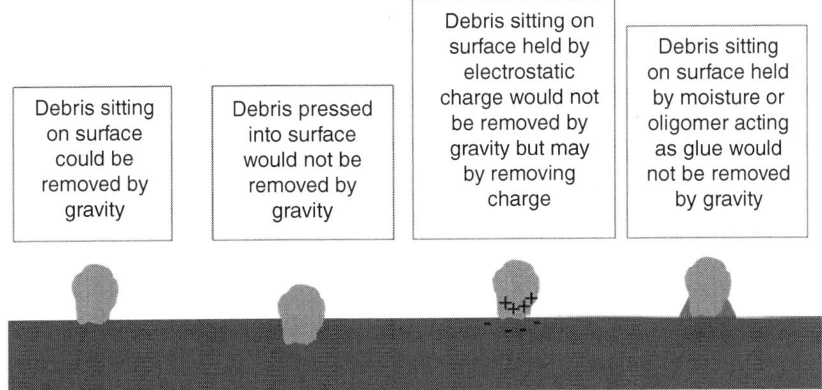

Figure 5.18 How debris might be attached to the substrate surface.

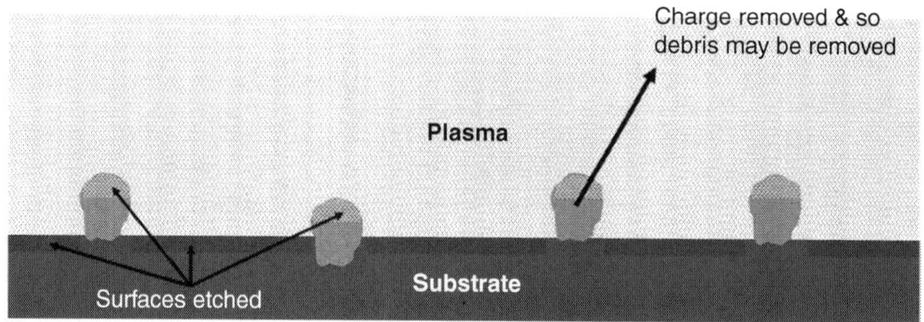

The plasma may allow any debris held by an electrostatic charge to be removed but for the rest the debris will be etched as is the substrate surface. As the debris may be microns in diameter the polymer surface could be expected to be etched by microns to have the plasma remove the debris by etching

Figure 5.19 Plasma is not a method that can remove debris by etching.

ture or oligomer or simply pressed into the surface. If we look at Fig. 5.19 where there is a plasma present we can see that the debris pressed into the surface or glued in place will not be affected by the plasma and will remain on the surface. If we consider that a plasma can etch material away we can see that etching away particles is not practical as if the debris were to be a micron in size we could easily etch away a micron of the polymer surface between the debris and cause significant substrate roughening. Thus we ought not to think of plasma as being a method of removing gross contamination such as debris but only for cleaning chemical contamination such as low molecular weight material such as additives or oligomers.

As mentioned above the plasma in corona, atmospheric or vacuum plasma treatments contains a significant proportion of ultraviolet (UV) light which enables photocatalytic processes to clean the film. This is shown schematically in Fig. 5.20. It is possible to use UV lamps along with oxygen to clean surfaces (78,79). This cleaning process is not often used with web processing but is possible. The process uses a high-powered quartz lamp as the source of UV energy. The UV energy is sufficient to speed up desorption of water and carbon dioxide from surfaces. The lamp is usually positioned where it can illuminate a large area and hence assist in desorption of these gases from the walls and vessel furniture during pump-down too.

Adding oxygen to the process can increase the effectiveness by increasing the production of ozone and atomic oxygen. This process can be particularly good at removing hydrocarbons.

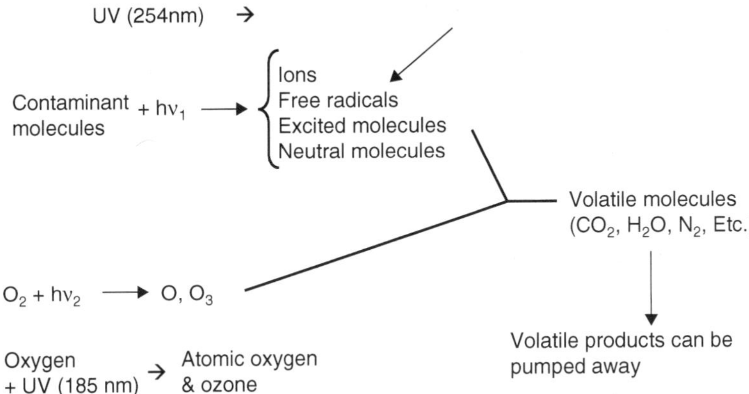

Figure. 5.20 A schematic of the conversion of UV light into species that can clean surfaces and in particular with oxygen can produce volatiles that can be pumped away.

5.4 Wetting and Adhesion

The easiest method of measuring the surface energy of polymer films uses a series of pens where each of the pens if filled with a specific liquid. Each liquid has a different but known surface energy. It is possible to buy a series of pens that can measure from low to high surface energies in dynes/cm and in 2 dyne/cm steps. For a polymer film of unknown surface energy one starts at one end of the range and each of the pens are wiped on a surface. Those pens that have a surface energy lower than the web will wet out the web surface and those pens with a surface energy higher than the web will ball up and not wet the surface. By checking the surface energy of the pen where there is a change from wetting to non-wetting behaviour it is possible to identify the surface energy of the web surface (80).

In Table 5.6 there is a list of liquids that are used for testing surface energy. As can be seen there is not a complete set of liquids at 2 dyne/cm intervals. To achieve the desired range of pens at 2 dyne/cm intervals the liquids are mixed in appropriate proportions. It is worth noting that these liquids can be volatile and so over time the pens can dry out. As the liquids have different levels of volatility the pens may not fully dry out but the dyne/cm value may become less accurate as one of the liquids dries out faster than the other and so the concentration changes, changing the surface tension of the remaining liquid. Hence care is required if this type of measurement is only carried

Table 5.6 The surface energy of some polymer film surfaces & the improvements following treatment on the primary three films used for barrier packaging along with the surface tension values of some commonly used test liquids.

Solids Polymers	**Untreated Surface free energy Dyne/cm (N/mm)**	**Treated Surface free energy Dyne/cm (N/mm)**	**Liquids**	**Surface Tension Dyne/cm (N/mm)**
Nylon	38–46		Water	72
Polyester	41–44	48–52	Glycerol	63
Cellulosics	42		Formamide	58
PVC, acrylic	39		Glycol	47
PMMA	30–36		Benzene, Toluene, Xylenes	29
polystyrene	38		Cyclohexane	25
polystyrene (low ionomer)	33			
L/HDPE	30–31	38–40	methyl, ethyl, iso-propyl - alcohols	22
Polypropylene	29–31	39–41		
PTFE	19–20		n-hexane	18

PVC = PolyVinyl Chloride.

PMMA = PolyMethyl Meth Acrylate.

L/HDPE = Low/High Density Poly Ethylene.

PTFE = Poly Tetra Flouro Ethylene.

out periodically. Using the alternative test method described below is more reliable when only occasional measurements are taken.

The alternative technique is slightly more time consuming (81). Using a syringe a drop of liquid, usually water, is dropped from near the surface onto the web (82). The water will either spread out and wet the surface or ball up depending on the surface energy of the web surface (83–85). From one side of the web a light is directed through the edge of the web and the side of the water drop and the image is projected onto a protractor. The angle that the water makes to the web can then be measured and used to calculate the surface energy of the web surface. The more the water spreads out the higher the web surface energy and hence the water is spreading out to try to minimise the combined water, air and web surface energies. Where the surface energy of the web surface is low the water will ball up, the more it approaches a sphere the lower the web surface energy. This is shown schematically in Fig. 5.21.

There are three measures of surface energy, the surface energy of the substrate, the surface tension of the liquid and the interfacial surface energy of where the liquid touches the substrate. The surface energy of a liquid is referred to as the surface tension with the units being the same as for the surface energy of a solid. Nature always tries to minimise the energy of any system and so when a liquid is dropped onto a solid surface the result will be the lowest combination of the three surface energies. So if the substrate has a very low surface energy the liquid is likely to tend towards a sphere with the liquid not wetting the surface. This is hydrophobic (water) or oleophobic (oil) repellent behaviour as shown on the left of Figure 5.21. If the substrate surface energy is maximised the liquid will tend towards wetting the surface as shown on the right of Fig. 5.21. Figure 5.22 shows how this is seen in practice for polypropylene films. The target is to achieve a

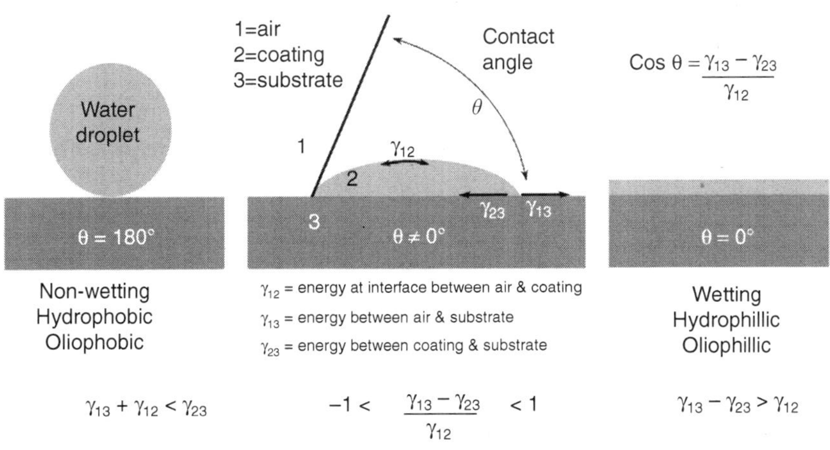

Ideally γ_{13} wants to be as high as possible with γ_{23} as low as possible
The system will want to equilibrate to the minimum total energy

Figure 5.21 A schematic showing the range from non-wetting to wetting behaviour.

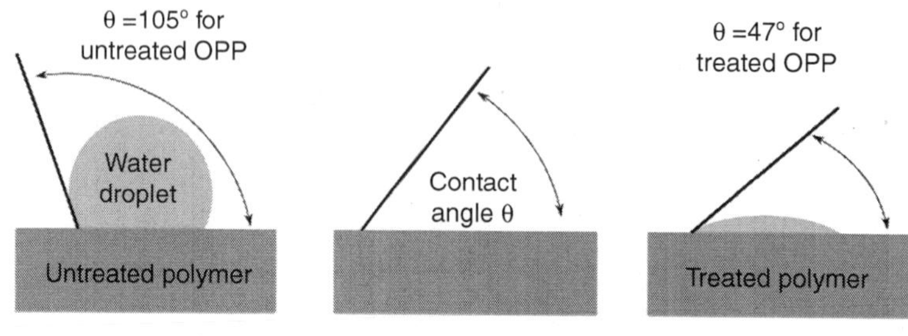

Figure 5.22 A schematic of the difference between treated and untreated polypropylene.

surface energy of between 3 dynes/cm and 10 dynes/cm greater than the surface tension of any liquid that is being applied to the surface. So if a polymer layer is being coated or an adhesive laminated to the surface knowing the surface energy or surface tension of the material being applied can provide a target surface energy that will enable good wetting and improved adhesion.

In the chapter on process there is a description of the nucleation and growth of vacuum deposited coatings. This section highlighted the need for the surface energy to be as high as possible to make sure the wetting of the growing material is as good as possible. By increasing the wetting the coating can grow closer to the Frank van der Merwe layer by layer growth method than the island or Volmer-Weber method. This leads to fewer grain boundaries and fewer pores making the barrier performance better. Also the coating will become continuous at a lower thickness and be smoother where the wetting is higher which is also more likely to provide a better barrier performance.

Non-wetting low surface energy surfaces leads to poor adhesion which can also lead to failures later such as delamination. The minute gaps between the metal and polymer can allow migration of oligomer, additives, air or water vapour to fill the gap. If air or water vapour fill the gap then rapid temperature changes can swell the air or water vapour and the pressure build up can cause the gap to spread and delamination to begin.

Hence wetting helps not only the immediate adhesion performance but also the longer term product performance.

5.5 Subbing or planarisation layers and over-coatings

Once it becomes clear that the coating can be affected so much by the surface quality of the substrate then one of the options for making improvements to the barrier performance of the coating is by controlling the substrate surface. There are the simple things that can be done such as cleaning of the surface both from gross contamination by dust/debris, using tacky rolls, as well as modifying chemical contamination by migrating additives or oligomer using plasma. However this has limitations as the tacky rolls cannot remove all the debris from the surface, fillers protruding from the surface will remain and plasma treatment designed to clean the surface can also roughen the surface.

To overcome these shortcomings subbing or planarisation layers are used, particularly where ultra barrier performance is required. These subbing or planarisation layers can be applied either outside or inside the vacuum system. The purpose of this layer is to smooth the surface and cover up as many of the surface defects as possible so that the vacuum deposited coating is as near perfect as possible making the barrier performance closer to the ideal. That is the thinking behind adding the layer but this coating still needs to be managed well to be of benefit.

If the coating is applied outside the vacuum system it too needs to be protected from becoming re-contaminated by debris. The coating process needs to be capable of improving the surface roughness over the original polymer surface. Also the coating thickness needs to be sufficient to cover the existing debris and defects. This can be a problem as debris on the surface will contain many particles greater than 10 microns in size which would require a coating in excess of 10 microns in thickness. This would significantly change the mechanical performance of the substrate as well as being expensive in material costs. Pre-cleaning the substrate by using one of the cleaning techniques such as the tacky roll to minimise the remaining debris size would enable a coating thickness of less than

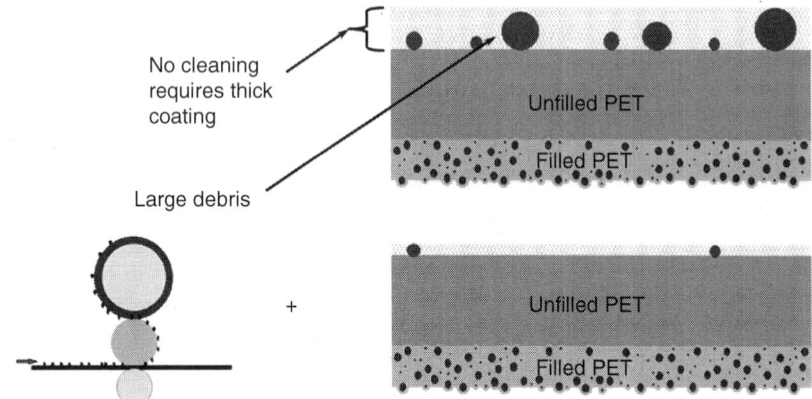

Figure 5.23 The minimisation of a subbing layer by pre-cleaning the polymer substrate.

1 micron to be required to cover up all of the remaining debris. This is shown schematically in Fig. 5.23.

Any coating applied to the substrate needs to have suitable adhesion to both the substrate and the metal coating vacuum deposited on top. The same considerations need to be given to the substrate before the subbing layer is applied and this may require a plasma treatment to prepare the surface for the coating. This coating needs to wet out the substrate well particularly if it is to cover up not only the debris but also any surface defects such as filler protruding through the polymer surface skin (86). This can also include covering up oligomers or if the planarising layer is added during the film manufacturing the planarising layer can provide a barrier to the oligomer exuding to the surface, reducing the problem (87). The coating once applied may need a settling time to form the smooth surface layer (88–91) that at least matches the substrate smoothness if not improves on it.

An alternative to applying the subbing coating outside the vacuum is to apply the coating inside the vacuum system. This technology began with the production of capacitors where a drum was spun around with alternating polymer and metal layers being applied to make up the capacitor (92).

This same process can be used on a roll-to-roll coating system where a single layer of polymer followed by the metal coating process can be done as a single pass. The polymer deposition process (93–95) is shown schematically in Fig. 5.24. The monomer is

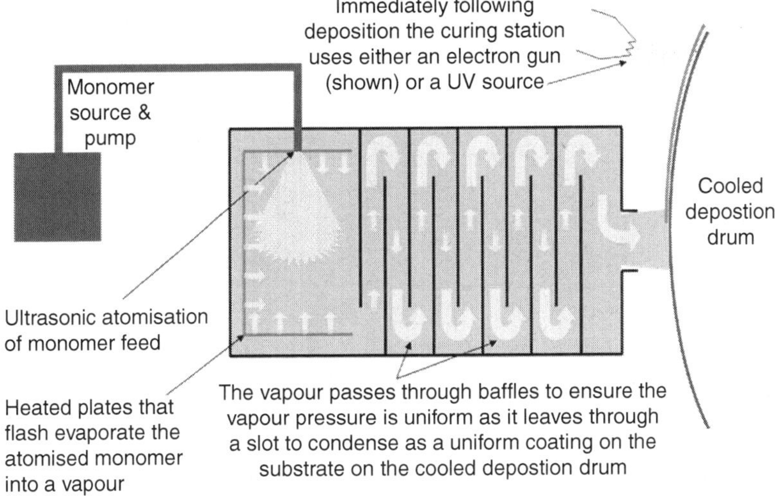

Figure 5.24 A schematic of the in-vacuum polymer deposition process.

metered and pumped and introduced into the vacuum system through an ultrasonic atomiser. The atomised monomer hits a hot plate where the micro-droplets are converted into a vapour which then via baffles, which help to even out any localised pressure variations, to an exit slot that directs the vapour onto the polymer substrate. The polymer substrate is on a cooled deposition drum and so the vapour condenses onto the polymer surface and is carried forward past a curing zone. The curing can be done by either electron beam or ultraviolet light. As the electron source and substrate are both under vacuum there is not the need to run the electron source at as high a voltage as when it is used to cure monomers at atmospheric pressure. The UV source does not need to be a lamp but can be from the UV light output from a plasma. This process has taken time to make robust for large industrial use. If you imagine the conversion of a liquid to a vapour in vacuum and the corresponding volume change it is not surprising that very small variations in the metered and pumped monomer may still be seen as a deposited coating thickness variation. Directing the vapour at the cooled polymer film is expected to give a high sticking coefficient but this is not necessarily so and it is often found that there is a significant amount of monomer that appears on most surfaces in the vacuum system and in the pumps. More recently experiments have been done to deposit the polymers by direct evaporation within the vacuum system (96) where many of these problems are avoided. The evaporation source does need to be very precisely controlled uniformly within a fraction of a degree to be able to maintain a stable evaporation rate. This type of source has been developed for molecular beam epitaxy deposition of semiconductors as well as more recently as slot sources for the deposition of organic light emitting devices (97–99). There is also no reason why printing or coating technology should not be used inside the vacuum system to deposit polymer coatings.

The original patents for this technology have run out but there are many subsequent patents that mean that this technology has not been exploited as much as one might expect considering the benefits.

Table 5.7 is from results presented on the benefits of using the polymer coating (100). As can be seen there are different options for the deposition of the polymer. The subbing layer is where the polymer is used directly onto the substrate and it can be seen that the number of pinholes is reduced by a factor of three. It can also be seen that not using a subbing layer but over-coating the aluminium gives an even better performance at approximately five times better. Using the polymer coating on top of the metal coating not only covers up the debris but also reduces the damage that can be done by the debris moving away as well as reducing pinholes from pick-off. Debris when moved can leave behind a pinhole and if it slides away can also leave a scratch. If the debris rolls away and if it does not disappear completely from the surface the same debris can do further damage picking off more metal as it goes. If the top surface is covered by a polymer this type of damage is prevented and so the number of pinholes is further reduced.

The improvement in barrier performance for the over-coating can be substantially more than for the subbing layer as the polymer

Table 5.7 Reduction of pinholes with polymer layers.

Sigma International - Benefits of polymer layers.	
Materials	**No. pinholes/unit area**
OPP + Al	1760
OPP + Polymer + Al	590
OPP + Al + Polymer	350

not only protects the coating but also is able to fill in any existing pinholes. The pinholes if filled with air will have the diffusion coefficient of air whereas if filled with polymer will have the much lower diffusion coefficient of the polymer and hence reducing the overall diffusion rate. This has been further improved by using a polymer with a nano-scale inorganic filler. The inorganic filler has a diffusion coefficient several orders of magnitude better than any polymer and so this further significantly reduces the diffusion coefficient through any pinholes or pores that are filled with the composite filled polymer coating (101).

This basic technology can also be used for both subbing and over-coating layers around the same vacuum deposited layer thus achieving the best out of both layers. This can also be extended to using multiple vacuum deposited inorganic layers. A pair of organic with inorganic coatings are sometimes referred to as a dyad and some ultra barrier layers have been developed by using multiple dyads (102–105).

References

1. Wagner Jnr. J.R. Ed. *'Multilayer flexible packaging'* Pub. Elsevier ISBN-13 978-0-8155-2021-4, 2009.
2. S.D. Ross et al. 'Isolation of a cyclic trimer from polyethylene terephthalate film' J. Poly. Sci., Vol.13. No.70. pp 406–407, April 1954.
3. I. Goodman & B.F. Nesbitt. 'The structures & reversible polymerisation of cyclic oligomers from poly(ethylene Teraphthalate)' *J. Poly. Sci.,* Vol. 48. No. 150. pp 423–433, December 1960.
4. A. Perovic & P.R. Sundararajan. 'Crystallisation of cyclic oligomers in commercial poly(ethylene terephthalate) films' *Poly. Bull.,* Vol. 6., pp 277–283, 1982.
5. Ito E. & Okajima S. 'Studies on cyclic tris(ethylene terephthalate) *Polymer* 12, (1971) pp 650–655.
6. Binns G.L. et al. 'Allotropy of cyclic tris(ethylene terephthalate)' *Polymer* Vol. 7, No 11, (1966) pp 583–585.
7. MacDonald W.A. 'Engineered films for display technologies' *J. Materials Chemistry* 14, 2004, pp 4–10.
8. DuPont Mylar Product Literature http//:www.dupont.com/packaging/products/films.
9. I. Goodman & B.F. Nesbitt. 'The structures & reversible polymerisation of cyclic oligomers from poly(ethylene Teraphthalate)' *Polymer,* Vol. 1. pp 384–396, 1960.

10. R. Giuffria. 'Microscopic studies of Mylar film and its low molecular weight extracts' *J. Polym. Sci.* Vol. 49. pp 427–436, 1961.
11. L.H. Peebles et al. 'Isolation & identification of the linear and cyclic oligomers of poly(ethylene terephthalate) & the mechanism of cyclic oligomer formation' *J. Polym. Sci.* A-1, Vol. 7. No. 2. pp 479–436, 1969.
12. M.T. de A. Freire et al. 'Thermal stability of polyethylene terephthalate (PET): Oligomer distribution & formation of volatiles' *Packaging Tech. Sci.* Vol. 12. No. 1. pp 29–36, 1999.
13. S.R. Shukla & K.S. Kulkarni. 'Estimation & characterisation of polyester oligomers' J. Appl. Polym. Sci. Vol. 74. pp 1987–1991, 1999.
14. Fayet G. et al. 'Effect of Anti-blocking Particles on Oxygen Transmission Rate of SiOx Barrier Coatings Deposited by PECVD on PET Films' 48th Annual Technical Conference Proceedings SVC, pp 237–240, 2005.
15. Rochat G. et al. 'Influence of substrate additives on the mechanical properties of ultrathin oxide coatings on poly(ethylene terephthalate)' *Surface & Coatings Technology* Vol. 200, Issue 7, 21 Dec. 2005, pp 2236–2242.
16. da Silva Sobrinho A. S. et al. 'Characterization of Defects in PECVD-SiO2 Coatings on PET by Confocal Microscopy' *Plasmas & Polymers* Vol. 3, No. 4, pp 231–247, Dec. 1998.
17. D.K. Davies. 'The generation and dissipation of static charge on dielectrics in a vacuum' Proc. Static Electrification Conf. pp 29–36, 1967.
18. 'The control of static charges on film' Information Note MX IN48 4th Edn. ICI Polyester Films 1986.
19. J.A. Griffiths & J.R. Lander. 'Packaging & static electricity' *Packaging* pp 29–36. February 1988.
20. H. Clow. 'Electrostatic charging of polymeric films in vacuum roll coaters' Proc. 3rd Intl. Conf. Vac. Web Coat., pp 97–107, 1989.
21. G. Rappe. 'The electrophoresis effect as a method to remove airborne & surface contaminants' Proc. 32nd Ann Tech Conf. Society Vacuum Coaters. pp 19–23, 1989.
22. L.E. Wetzel. 'The invisible enemy – Identification & control of airborne contaminants in the manufacturing environment' Proc. 33rd Ann.Tech.Conf. SVC. pp 262–268, 1990.
23. E.H.H. Jamison & A.H.Windle. 'Structure and oxygen barrier properties of metallized polymer films' *J. Matls. Sci.* Vol. 18, pp 64–80, 1983.
24. Weiss, J. 'Parameters that influence the barrier properties of metallized polyester and polypropylene films' *Thin Solid Films*, Vol. 204, issue 1, pp 203–216, 1991.
25. Taylor A. 'Practical solutions to web heating problems' Proc. 7th Intl. Conf. Vac. Web Coat pp 107–119, 1993.

26. McCann M. J. et al. 'Buckling or wrinkling of thin webs off a drum' Proc. 47th Ann Tech. Conf. SVC 2004, pp 638–643.
27. da Silva Sobrinho A.S. et al. 'Defect-Permeation Correlation for Ultrathin Transparent Barrier Coatings on Polymers' *J. Vac. Sci. Technol., A* 18(1), American Vacuum Society, pp 149–157, 2000.
28. da Silva Sobrinho A.S, 'A Study of Defects in Ultra-Thin Transparent Coatings on Polymers' *Surf. Coat. Technol.* 116–119 (1999) 1204.
29. da Silva Sobrinho A.S. et al. 'A Study of Defects in Ultra-Thin Transparent Coatings on Polymers', *Applied Physics A, Materials Science & Processing*, 68, pp 103–105, 1999.
30. Hamilton S. 'Yield improvement in metallizing lines by using contact cleaning technology' Proc. 51st Ann Tech. Conf. SVC 2008 pp 778–780.
31. 'Web cleaning efficiency using contact cleaning roll (CCRs).' Down loaded from http://www.polymagtek.com/pdf/dirt_rpt.pdf.
32. Hamilton S. 'Advances in Contact Cleaning for Yield Improvement' Proc. AIMCAL Fall Annual Technical Conference 2008.
33. Attachment A – nip efficiency test Down loaded from http://www.polymagtek.com/pdf/dirt_rpt.pdf.
34. Weiss H. 'Web cleaners can eliminate dust-adhesion problems' Paper, film & foil converter. Nov. 1991 pp 91–94.
35. Owens D.K. 'The mechanism of corona & ultraviolet light induced self adhesion of poly(ethylene terephthalate) film.' *J. Appl. Polymer Sci.* Vol. 19, 1975 pp 3315–3326.
36. Briggs D. et al. 'X-ray photoelectron spectroscopy studies of polymer surfaces: Part 3, flame treatment of polyethylene.' *J. Materials Sci.* Vol. 14, 1979 pp 1344–1348.
37. Briggs D. et al. 'Surface modification of polyethylene terephthalate by electrical discharge treatment.' *Polymer* Vol. 21, 1980 pp 895.
38. Briggs D. 'Surface treatments for polyolefins.' Chapter 9 Ed. Brewis D.M. 'Surface analysis & pretreatment of plastics & metals. Pub. Applied Science 1982.
39. Briggs D. et al. 'Electrical discharge treatment of polypropylene film.' *Polymer* Vol. 24, 1983 pp 47.
40. Podhajny R.M. 'Corona treatment of polymeric films.' *J. of Plastic film & sheeting* Vol. 4, 1987 pp 177–188.
41. Mittal K.L. 'Surface cleaning: recent developments & needs.' Proc. 34th Ann. Tech. Conf. SVC 1991 pp 13–17.
42. DiGiacomo J.D. 'Advanced technology flame plasma surface treating system.' Proc. 36th Ann. Tech. Conf. SVC 1993 pp 356–361.
43. Strobel M. et al. 'Analysis of air-corona-treated polypropylene and poly(ethylene terephthalate) films by contact-angle measurements & x-ray photoelectron spectroscopy.' pp 493–508 In *'Contact angle, wetability & adhesion.'* Ed. Mittal K.L. Pub. VSP Utrecht 1993.

44. Mount III E.M. & Bishop C.A. Eds. *'Metallizing technical reference.'* 4th Edn. 2007 Pub. The Association of Industrial Metallizers, Coaters & Laminators.
45. Barankova H. & Bardos L. 'Cold atmospheric plasma sources for surface treatment.' Proc. 46th Ann. Tech. Conf. SVC 2003 pp 427–430.
46. Goodwin A. et al. 'Atmospheric pressure plasma technologies for surface modification of polymers.' AIMCAL 2003 Proc. 17th Vacuum Web Coating Conf. Session 2B, Section 2, Paper 1.
47. Korzec D. et al. 'Atmospheric pressure plasma source for wire cleaning.' Proc. 47th Ann. Tech. Conf. SVC 2004 pp 455–450.
48. A.T. Devine & M.J. Bodnar 'Effects of various surface treatments on adhesive bonding of polyethylene.' *Adhesives Age* 12(5) p35 May 1969.
49. Hall J.R. et al. 'Activated gas plasma surface treatment of polymers for adhesive bonding.' *J. Appl. Poly. Sci.* Vol. 13, 1969 pp 2085–2096.
50. Holland L. 'Substrate treatment & film deposition in ionized & activated gases.' *Thin Solid Films* Vol. 27, 1975 pp 185–203.
51. Schonhorn H. et al. 'Surface modification of polymers & practical adhesion.' *Polymer Eng. & Sci.* Vol. 17, No. 7, 1977 pp 440–449.
52. Holland L. 'Glow discharge excitation & surface treatment in low-pressure plasmas.' Inst. Phys. Conf. Ser. No. 54, Chapter 6, 1980 pp 220–228.
53. Liston E.M. 'Plasma treatment for improving bonding: a review.' *J. Adhesion.* Vol. 30, 1989 pp 199–218.
54. Schut J.H. 'Plasma treatment: the better bond.' Plastics Technol. Oct 1992 pp 64–68.
55. Liston E.M. et al. 'Plasma surface modification of polymers for improved adhesion: a critical review.' *J. Adhesion Sci. Technol.* Vol. 7, 1993 pp 1091–1127.
56. Smithson R.L.W. 'Nucleation & growth of gold films on bare & modified polymer surfaces.' Proc. 38th Ann Tech. Conf. SVC 1995 pp 72–79.
57. Kaplan S.L. 'Cold gas plasma treatment for re-engineering films.' Paper, film & foil converter.' June 1997 pp 70–74.
58. d'Agosino R. et al. 'Plasma processing of polymers.' NATO ASI series Vol. 36, Kluwer Netherlands 1997.
59. Kaplan S.L. 'Cold gas plasma treatment of films, webs & fabrics.' Proc. 41st Ann Tech. Conf. SVC 1998 pp 345–348.
60. Nickerson R. 'Plasma surface modification for cleaning & adhesion.' Vac. Technol. & Coat. June 2000 pp 56–60.
61. Mattox D.M. 'Surface preparation: plasma cleaning.' *Vacuum Technol. & Coating* Vol. 1, No. 6, Aug 2000 pp 14.
62. Briggs D et al. 'Surface modification of polyethylene terephthalate by electric discharge treatment' *Polymer,* 21, pp 859–900, 1980.

63. Brewis D.M & Briggs D. 'Adhesion to polyethylene & polypropylene' *Polymer*, 22, pp 7–16, 1981.
64. Garbassi F. et al. 'Surface effect of flame treatments on polypropylene' *J. Material Science*, 22, pp 1450–1456, 1987.
65. Ayres R.L. & Shofner D.L. 'Preparing polyolefin surfaces for inks and adhesives' *SPE Journal*, 28, No. 12, pp 51–55, Dec 1972.
66. Lawson D & Greig S. 'Bare roll treaters versus covered roll treaters: Make the right choice' TAPPI Polymers, Laminations & Coatings Conference. Toronto, Ontario 24th – 28th Aug 1997 Conference proceedings Book 2. pp 681–693, 1997.
67. Mount III. E.M & Bishop C.A. Eds. 'Section 7: Surface treatments of webs for aluminium vapour deposition' Metallizing Technical Reference 4th Edn. Pub. AIMCAL. pp 96–110, 2007.
68. Thomas M et al. 'Coating with Atmospheric Pressure Plasma Processes: From Large Area to μ-Structures' 49th Annual Technical Conference Proceedings SVC, pp 394–399, 2006.
69. Twomey B et al. 'Evaluation of the Mechanical Behaviour of Nanometre-thick Coatings Deposited Using an Atmospheric Pressure Plasma System' 50th Annual Technical Conference Proceedings SVC, pp 58–63, 2007
70. Weltmann K-D et al. 'Plasma decontamination at atmospheric pressure – Basics & applications' 51st Annual Technical Conference Proceedings SVC, pp 605–611, 2008.
71. Song S. & Placido F. 'Effect of Gas Release from Polymer Surfaces During Microwave Plasma Pre-cleaning on the Adhesion of Metal Films' 51st Annual Technical Conference Proceedings SVC, pp 574–578, 2008.
72. Ling A.O. et al. 'The Aging Study on Polyethylene Terephthalate with Surface Modification by Water Vapour Plasma' *Plasma Science & Technology*, Vo1. 3, No. 3, pp 761–764, (2001).
73. Papakonstantinou D. et al. 'Improved Surface Energy Analysis for Plasma Treated PET Films', *Plasma Process. Polym.* 2007, 4, S1057–S1062.
74. C.R. Morelock, Y. Htet, L. Wright and E.C. Culberson. "AFM Studies of Corona-treated, Biaxially Oriented PET Film," Converting Magazine, December 2007, as Presented at AIMCAL's Fall Technical Conference, 2007.
75. M. Strobel, C.S. Lyons, J.M. Strobel and R.S. Kapaun. "Analysis of air-corona-treated polypropylene and poly(ethylene terephthalate) films by contact-angle measurements and X-ray photoelectron spectroscopy". *J. Adhesion Sci. Technol.* Vol. 6. No. 4, pp 429–443 (1992).
76. Y. DePuydt, P. Bertrand and Y. Novis. "Surface Analyses of Corona-Treated Poly(ethylene terephthalate), *Br. Polym. J*, Vol. 21, No. 2 pp 141–146, 1989.

77. Prosyčevas I. et al. 'Polyethylene terephthalate film surface relaxation after plasma treatment' ISSN 1392–1320 *Materials Sci.* (MEDŽIAGOTYRA). Vol. 9, No. 4. 2003, 355–357.
78. Vig J.R. and LeBus J.W. 'UV/Ozone cleaning of surfaces' *IEEE Trans. Parts, Hybrids and Packag.*, Vol. PHP-12, 1976 pp 365–370.
79. Lasky E. 'The UV/ozone process for ultra-cleaning before vacuum deposition' 34th Annual Technical Conference Proceedings SVC, 1991, pp 374–377.
80. ASTM 2578-99a. 'Standard test method for wetting tension of polyethylene & polypropylene films.' 1999 (or ASTM 2578-84 1984 same title) American Society for Testing & Materials.
81. Hansen C.M. 'Characterization of surfaces by spreading liquids.' *J. of Paint Technology* Vol. 42, No. 550, Nov. 1970 pp 660–664.
82. ASTM D 5946-96. 'Standard test method for corona-treated polymer films using water contact angle measurements.' American Society for Testing & Materials. Approved 1996. Published June 1996.
83. Good R.J. 'Contact angle, wetting, & adhesion: a critical review.' pp 3–36 In '*Contact angle, wetability & adhesion.*' Ed. Mittal K.L. Pub. VSP Utrecht 1993.
84. Padday J.F. 'Spreading, wetting & contact angles.' pp 97–108 In '*Contact angle, wetability & adhesion.*' Ed. Mittal K.L. Pub. VSP Utrecht 1993.
85. Wetterman B. 'Contact angles measure component cleanliness.' Precision Clean. Oct 1997 pp 21–24.
86. Adam R. et al. 'Optimising Polyester Films For Flexible Electronic Applications' Proc. AIMCAL Fall Annual Technical Conference 2008.
87. MacDonald W.A. 'Engineered films for display technologies' *J. Materials Chemistry* 14, pp 4–10, 2004.
88. Orchard S.E. 'On surface levelling in viscous liquids & gels.' *Appl. Sci. Res. Ser. A*, Vol. 11, 1962 pp 51.
89. Satas D. '*Coatings technology handbook.*' 1991 Pub. Marcel Dekker Inc. ISBN 0-8247-8410-3.
90. Anshus B.E. 'The levelling process in polymer powder paints – a three dimensional approach.' *A. Chem. Soc. Div. Org. Coat. Plast. Chem.* Pap. Vol. 33, 1973 pp 493.
91. Kistler S.F. & Schweizer P.M. Eds. '*Liquid film coating – scientific principles & their technological implications.* Pub. Chapman & Hall 1997 ISBN 0-412-06481-2.
92. US 4,508,049. 'Method & a device for the production of electrical components, in particular laminated capacitors.' Behn R. et al. 1985 US 4,378,382 'Method & apparatus for producing laminated capacitors.' Behn R. 1983 US 4,490,774 'Capacitors coataining polyfunctional acrylate polymers.' Olson D. 1984 US 4,499,520 'Capacitor with dielectric comprising poly-functional acrylate polymer & method of making.'

Cichanowski S. 1985 US 4,513,349 'Acrylate-containing mixed ester monomers & polymers thereof useful as capacitor dielectrics.' Olson D. 1985 US 4,515,931 'Polyfunctional acrylate monomers & polymers thereof useful as capacitor dielectrics.' Olson D. 1985 US 4,533,710 '1,2-Alkanediol diacrylate monomers & polymers thereof useful as capacitor dielectrics.' Olson D. 1985 US 4,586,111 'Capacitor with dielectric comprising a polymer of polyacrylate polyether pre-polymer. Cichanowski S. 1986 US 4,842,893 'High speed process for coating substrates.' Yalizis A. & Shaw D. 1989 US 4,647,818 'Nonthermionic hollow anode gas discharge electron beam source.' Ham M. 1987.

93. Affinito J.D. et al. 'A new hybrid deposition process, combining PML & PECVD, for high rate plasma polymerisation of low vapor pressure, & solid, monomer precursors.' Proc. 42nd Ann Tech. Conf. SVC 1999 pp 102–106.
94. Shaw D.M. & Langlois M.G. 'A new high speed process for vapor deposition arcylate thin films: an update.'Proc. 36th Ann Tech. Conf. SVC 1993 pp 348–352.
95. Yalizis A. & Mikhael M. 'UV versus electron beam radiation curing of vacuum deposited polymer levelling coatings for high barrier applications.' Proc. 47th Ann Tech. Conf. SVC 2004 pp 600–605.
96. Affinito J. 'Polymer film deposition by a new vacuum process. Proc. 45th Ann Tech. Conf. SVC 2002 pp 425–439.
97. Van Slyke S. et al. 'Linear source deposition of organic layers for full-color OLED.' Soc. Info Disp. Digest 2002 pp 886–889.
98. US 6,237,529 patent Spahn R.G. 'Source for thermal physical vapor deposition of organic electroluminescent layers' Pub. 2001.
99. Hoffmann U. et al. 'New in-line machine concept for OLED manufacturing.' Soc. Info Disp. Digest 2002 pp 691–693.
100. Yalizis A. et al. 'Barrier degradation in aluminium metallized polypropylene films' Proc. 40th Ann Tech. Conf. SVC 1997, pp 371–375.
101. Ramadas S. et al. 'Nanoparticulate Barrier Films and Gas Permeation Measurement Techniques for Thin Film Solar and Display Applications' Proc. AIMCAL Annual Technical Conference 2008.
102. Kapoor S. et al. 'Pilot Production of Ultrabarrier Substrate for Flexible Displays Proc. 49th Ann Tech. Conf. SVC 2006, pp 625–631.
103. Affinito J.D. et al. 'A new method for fabricating transparent barrier layers' *Thin Solid Films* Vols 290–291 Dec 1996, pp 63–67.
104. Affinito J.D. et al. 'PML/oxide/PML barrier layer performance differences arising from use of UV or electron beam polymerization of the PML layers' *Thin Solid Films* Vols 308–309, Oct 1997, pp 19–25.
105. Sung Jin Bae et al. 'Enhancement of Barrier Properties Using Ultrathin Hybrid Passivation Layer for Organic Light Emitting Diodes' *Jpn. J. Appl. Phys.* 45, 2006, pp 5970–5973.

6

Vacuum Deposition Processing

The key step in the process is the nucleation and growth of the depositing coating but to only look at this part of the process is to neglect other aspects that will affect the barrier performance of the deposited coating. So included in this section will be some other factors that must be taken into consideration such as the heat load the polymer film will see during the deposition process and beyond as well as winding webs in vacuum which will often relate to the incoming polymer substrate. In looking at how defects are created and can be minimized it gives the opportunity to look at some of the other factors that can affect the coating barrier performance such as system cleaning.

6.1 Nucleation, Growth and Modification

As nucleation and growth is such an important process there is a large body of work that has been done to try to understand the mechanisms involved (1–12). Much of the work has been done on idealized substrates such as glass slides or freshly cleaved surfaces, crystals or mica but the principles these experiments highlighted are generally applicable to most substrates.

The starting point for looking at the nucleation of a coating is simple, we have our polymer substrate that is passing over a deposition source that is providing atoms that will hit the substrate and the accumulation of these atoms will produce our coating. Not all coatings are the same and the differences can often be traced back to how this deposition process first starts with the first of the atoms reaching the surface. If we look at when the first atoms hit the surface there are several things that can happen. The atom can hit the surface and bounce back off the surface, or the atom can hit the surface and remain motionless, or the atom can hit the surface and move around on the surface for a while and then leave the surface, or the atom can initially move around on the surface before finally stopping somewhere. Every atom that hits the surface and stays on the surface is said to have a sticking coefficient of one, whereas if none of the atoms stay on the surface but all evaporate back off the surface is said to have a sticking coefficient of zero. Vacuum depositing barrier coatings onto polymer webs generally is at the more favourable end of this range with sticking coefficients close to one. What happens on the surface relates to the energy that is available. If the substrate is heated the behaviour of the atom arriving on the surface can be change. The atom arrives at the surface with some energy and it moves around until it finds a spot where the total energy of the surface with the atom on it can be minimized, this is known as an energy-well. This can be a defect on the surface such as a scratch where the atom will move to the bottom of the scratch where it may be in contact with two surfaces rather than one. Examples of possible bonding sites is given in Fig. 6.1.

Or if there is some debris on the surface the atom can touch both the debris and the substrate and so reduce the total energy. So if we look at the surface after there has been some atoms deposited the atoms

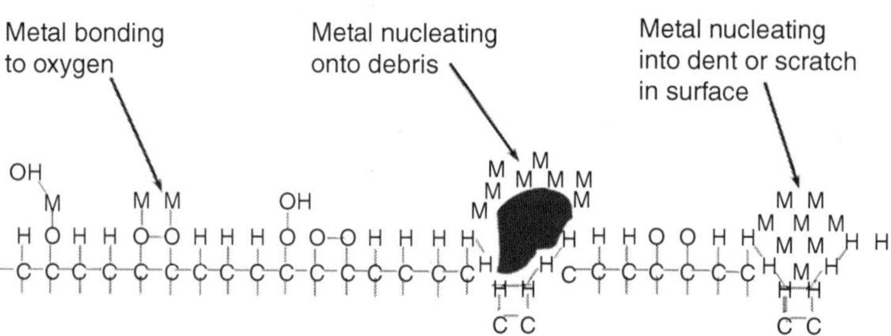

Figure 6.1 A schematic of possible bonding sites showing how defects can affect nucleation.

will mostly be located at various surface defects. Heating the substrate will provide more energy and so allow the atoms to move further and more quickly than if the substrate was at a lower temperature. This is an important factor as if the deposition rate is slow and the substrate temperature is high it is possible to have relatively few nucleation sites which will result in a large crystal size in the grown coating with fewer grain boundaries. Conversely if the substrate temperature is low and the deposition rate is high the resultant coating will have a much finer crystal size with many more grain boundaries.

Additional atoms will still be arriving from our deposition source even whilst the first atoms are still moving around on the surface. The newly arrived atoms will follow the same process of moving around on the surface until they find a place where they can reduce their energy. This may be at a defect or they may collide with another atom moving around on the surface. If they hit another atom they can stick together and in doing so will lose energy and will stop moving. Atoms that have stopped moving may also have other atoms hit them and so the growth process begins. As these nuclei grow they also increase in area and so some atoms arriving may land on top of these atoms and the thickness of the nuclei grows too. So we have a period of growth of existing nuclei but also in the gaps between nuclei there can still be more atoms arriving that will form more new nuclei. Thus we can see that there will be many nuclei that can be at different points in the nucleation and growth process. Figure 6.2 shows the nucleation and growth as a series of steps and this is shown schematically in Fig. 6.3 and the same progression is shown for gold on glass in Fig. 6.4.

In the schematics 3 & 4 in Fig. 6.3 there has been included a scratch to highlight how the atoms will congregate around such defects. The nuclei will continue growing and more nuclei will form

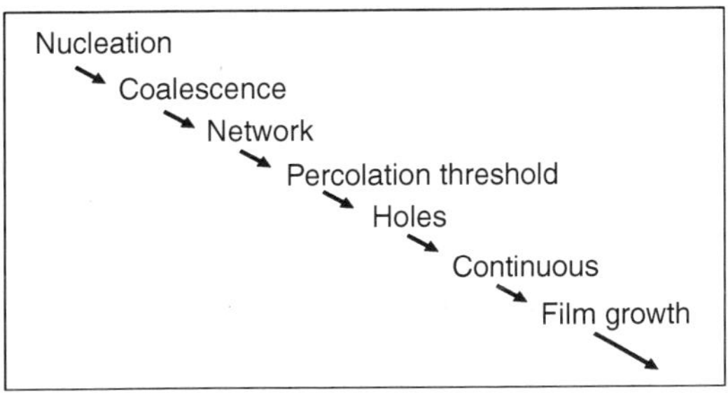

Figure 6.2 The general sequence of events representing nucleation and growth.

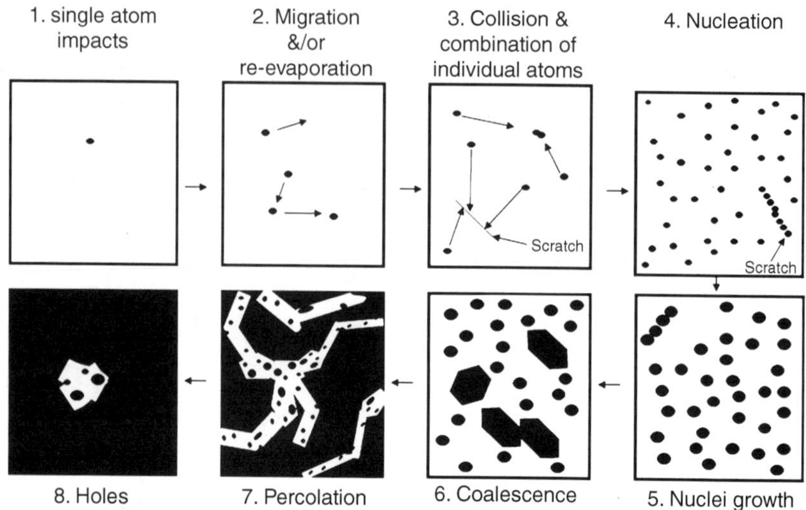

Figure 6.3 A schematic of the significant steps in nucleation and growth of coatings.

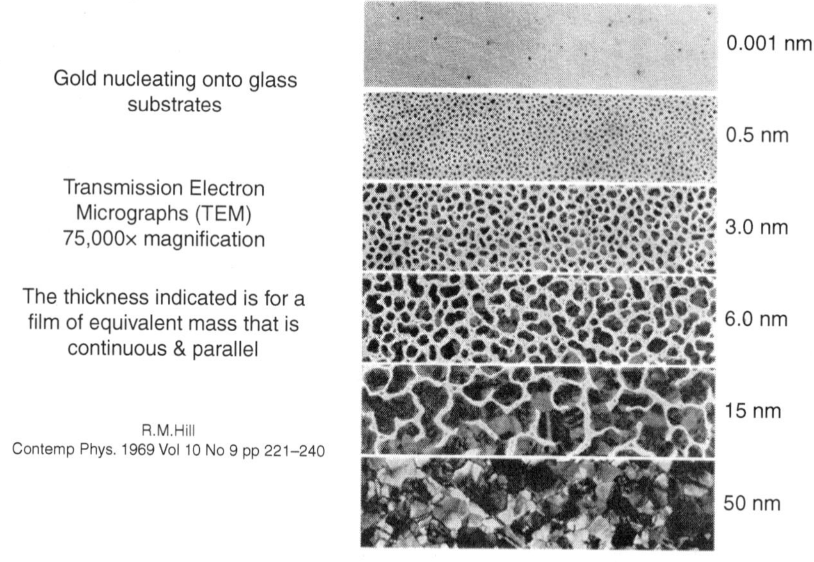

Figure 6.4 Micrographs showing the progression of nucleation and growth of gold on glass (8).

as the deposition process continues. As these islands keep growing some will grow close to near neighbours and when islands touch they will coalesce. The two islands may not have the atoms in the same orientation but during coalescence the atoms will rearrange

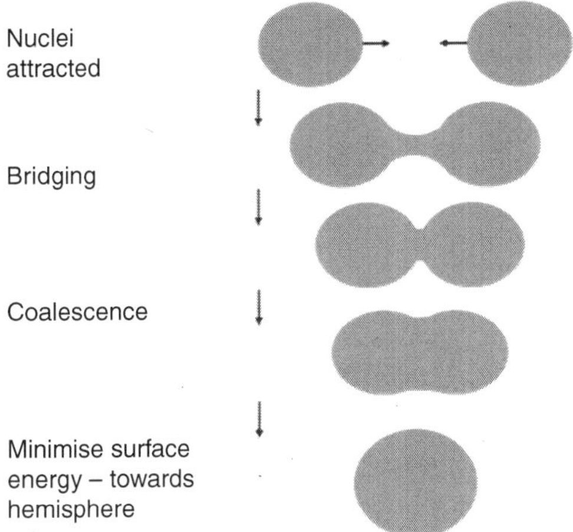

Figure 6.5 A schematic of the process of coalescence.

themselves so they all become one crystal. Often there is enough energy to make some of the islands mobile such that they do not have to touch each other to start the coalescence but only need to be near enough and the islands will move towards each other (13,14). This is shown schematically in Fig. 6.5. This process will continue with some islands getting bigger by more atoms adding to them and others growing by coalescence. Where islands move and they leave behind bare substrate where more nuclei can form. This growth will continue and progressively more of the surface will be covered with islands. A point will be reached where if connections are made to each side of the polymer film it will be possible to measure a resistivity. This point where the first measurement can be made as there is a complete conduction path from one side of the film to the other is called the percolation threshold (15,16).

Beyond the percolation threshold the addition of more atoms will increase the thickness of existing islands or interconnecting surface and progressively fill in any channels and holes between the coated material. Filling in the holes can take quite a long time and the thickest part of the coating can be very thick by comparison to the extremely thin coating at the bottom of the last hole to be filled in. A schematic of this thickness variation is shown in Fig. 6.6. This variation is quite important in regard to the barrier performance. Many of the opaque coatings used in barrier applications are specified by other parameters than the barrier performance. This might

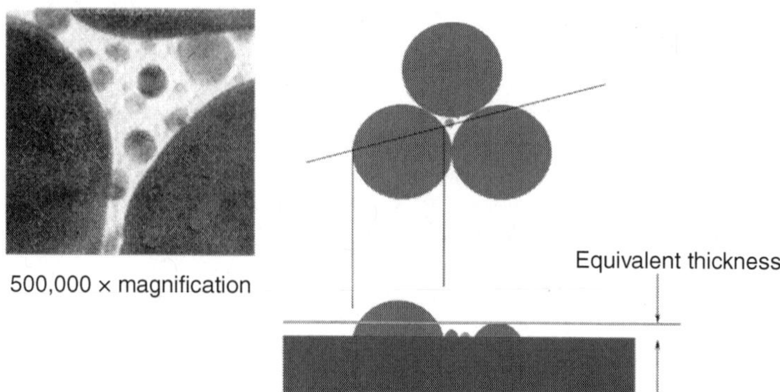

Figure 6.6 A high magnification micrograph of where three islands meet showing nuclei within the gap and a schematic of the thickness along a line drawn through a couple of the larger islands.

be optical density or coating thickness however it is noticeable that coatings with the same optical density may be of very different thickness. When a coating thickness is quoted it is important to note that this thickness is likely to be an average thickness. If an eddy current monitor is used it will be measuring the average resistivity over an area and similarly the measurement of reflectance, transmittance or the optical density will be the average performance over an area. Within this area there will be some parts that are thicker than average and others that are thinner than average as shown in Fig. 6.6.

In the micrograph in Fig. 6.6 the high magnification allows us to see the three large islands with a number of nuclei in the gap between. These small nuclei are attracted by the large mass of the islands and the nuclei form and move and are absorbed with new nuclei being formed in the gap all the time. Eventually the islands will touch and there will probably be a small pore left behind that is very difficult to completely block off. It is this type of pore that reduces the barrier performance from the theoretical to the more modest levels normally quoted.

The size and shape of the islands will determine the coating surface roughness, the optical performance as well as the resistivity for metallic coatings and the barrier performance. Hence anything we can do to the polymer surface will enable us to control the nucleation and growth of the coatings and so control the final performance. This starts with the surface roughness, chemistry and surface energy of the polymer film surface as well as the energy that we can add to the surface during deposition. In the previous chapter there was a description of the effects of surface energy and wetting. The aim is to increase the surface energy of the polymer substrate to maximise the wetting of the depositing coating. This is shown schematically in Fig. 6.7 which shows how if there is island growth on a low surface energy substrate the final surface will be rougher than if there is good wetting which encourages the islands to grow wider and flatter in aspect. There are three growth types that have been described (17,18) including how defects can affect growth and these three types are shown schematically in Fig. 6.8. Coatings that are deposited onto polymer films grow by Volmer-Weber process which is by the nucleation and growth of islands as described above. However if you look at the difference in aspect ratio of islands between the non-wetting and wetting surfaces the non-wetting surface follows the classic Volmer-Weber growth whereas the wetting surface whilst

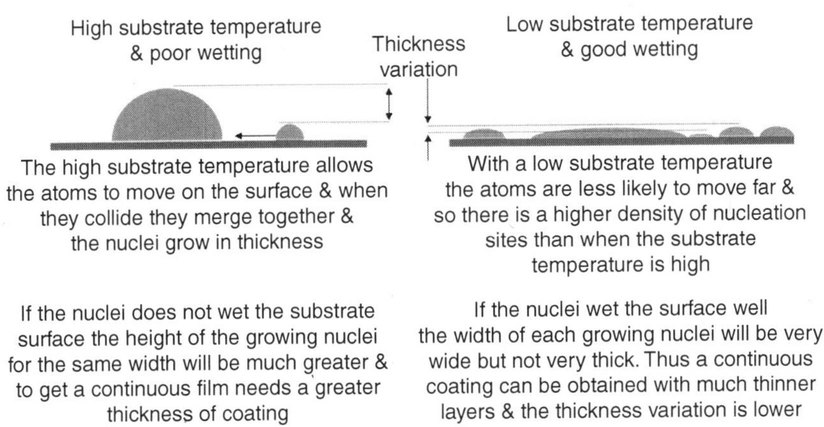

Figure 6.7 A schematic of how surface energy and wetting can affect the nucleation and growth and how this affects the coating thickness variation and surface roughness.

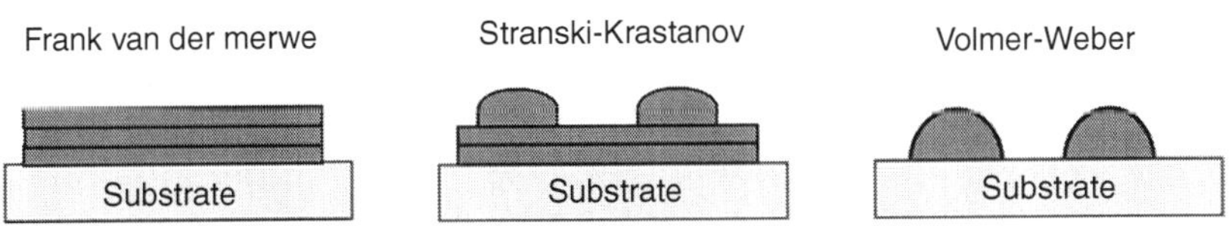

Figure 6.8 Schematics of the standard three types of growth mechanisms.

still following this growth method also shows some characteristics of the Frank van der Merwe type growth. The higher the surface energy and more the wetting the wider and thinner will be the islands which more approximate this type of growth. It will never be really Frank van der Merwe type growth because this requires a close match between the crystal structure of the substrate and coating allowing a good lattice match.

The nuclei and islands that grow from them will be random in the crystal orientation and the subsequent growth planes. Figure 6.9 shows how this can occur. As the atoms collect together they can form different patterns. If we take just four atoms they can either make up a square with the atoms being at the four corners or if two atoms are moved slightly they can fit closer together as shown in the left hand two schematics in Fig. 6.9.

If the closer fitting atoms have some more atoms added a hexagonal pattern can be made. If the atoms that sit on the corners of a square are moved apart slightly another atom could fit into the centre and this face centred cubic structure has a different atom spacing with the distance between layers of atoms being smaller than if they were in the hexagonal pattern which in turn is smaller than if a simple cubic structure was formed. The type of crystal growth is material dependent but even for the same crystal type there are different atom spacing dependent upon the crystal orientation. If we collect eight atoms to form a cube we can draw a plane across the face of the cube or across the opposite corners and the distance between

If we take the atoms but instead of having the spheres touch at the corners of a cube but have them pack as closely as possible they will form the hexagonal pattern as shown schematically in the middle right. The distance between planes has been reduced from 'a' in the simple cubic lattice to 'b' in the hexagonal lattice

On the far right is another lattice structure where again the distance between planes has been reduced from dimension 'a' down to dimension 'c'. This new structure has an atom in the centre of each face of the simple cubic lattice & hence it is called a face centre cubic (FCC) lattice

Figure 6.9 Some of the basic lattice or crystal patterns and the atom spacing differences.

each of these layers will vary. The basics of crystal plane numbering is shown if Fig. 6.10. The difference in growth planes and spacing is shown schematically in Fig. 6.11. This becomes important as the crystals grow. If there are two islands growing side by side but with different orientations then for the same number of atoms depositing the one with the wider atom spacing will grow taller faster than the one with the smaller atom spacing. This is shown schematically on the left of Fig. 6.11.

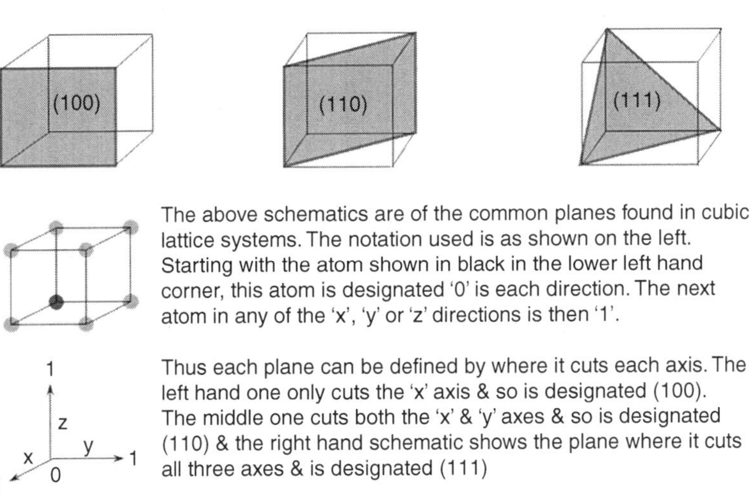

Figure 6.10 A schematic of the nomenclature of the different planes in crystal growth.

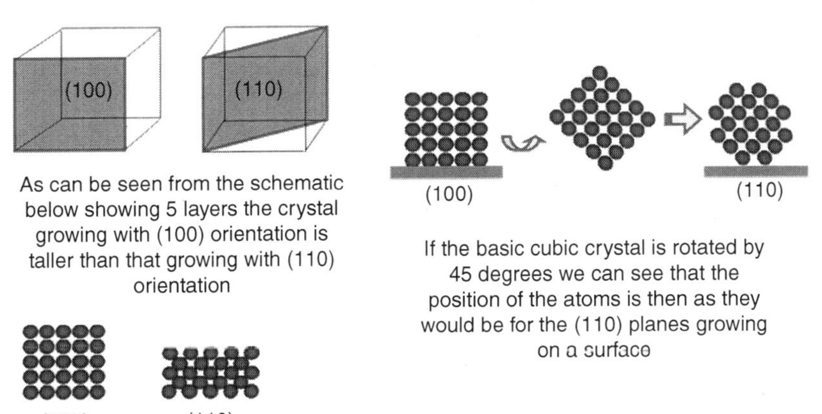

Figure 6.11 A schematic showing how orientation can affect the speed of crystal growth.

As the crystals or grains can grow at different speeds this also contributes to surface roughening. This roughening is less noticeable for very thin coatings but becomes more noticeable as the thickness increases. In aluminium coatings deposited onto polymer films the expectation is that the coating will be a mirror coating. This is true for most thin coatings but if the thickness is increased the surface can appear 'milky' where there is a grey or whitish tinge to the surface and the reflectivity decreases with a dullness beginning to appear. Further increasing the thickness will increase this dull or matte appearance. Sometimes the colour can appear to have a yellowish tinge and this can be related to oxidation of the rough surface.

Figure 6.12 shows what happens to coatings if they are grown into thick layers. Many people have different ideas of what constitutes a thick or a thin coating. I shall use 1 micron as my cut off point with coatings less than 1 micron being thin coatings and those over 1 micron being thick coatings. In other industries such as the machine tool industry where they may grow coatings to greater than 50 microns they would regard our thin coatings of up to 1 micron as part of the nucleation process. So it is worth being

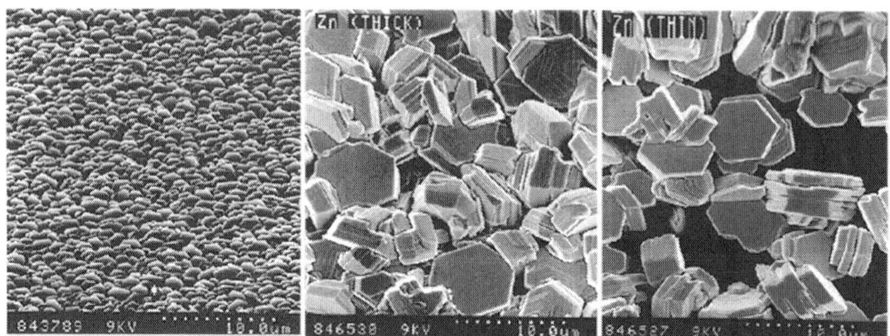

The above micrographs are all taken at the same magnification

The left hand micrograph shows the faceting of an evaporated magnesium coating of approximately 10 microns thickness. Note the crystal size is much smaller than for the micrographs of the zinc coating deposited using the same system.

On the right is a thin film that shows the zinc has formed well defined crystals in characteristic shapes. Also clearly seen are large clear areas without any coating highlighting the zinc has not wet the substrate. The middle micrograph shows that as the zinc thickness increases the holes are eventually filled in, but the coating remains porous & very rough.

Figure 6.12 Micrographs of different materials showing the effects of different growth patterns.

careful when reading about deposition and growth of coatings that their definition of nucleation, thin and thick coatings are compatible. Sometimes they can refer to a nucleation layer that may be thicker than any of our complete barrier coatings.

Figure 6.13 shows the growth of crystals showing how the different growth rates can cause shadowing where the higher dominant crystals will spread and further starve the lower crystals. This results in the crystal size being bigger on the surface than at the interface. This type of shadowing growth results in a columnar structure. This is not particularly good for a barrier coating as the gaps between columns can be porous. Included in the schematic in the centre is what happens if there is an additional plasma used during the deposition process. The bombardment by the energetic species in the plasma can knock off any loose atoms and so the adhesion increases also it is possible for atoms to cascade down and densify the coating. This can be disruptive to the crystal growth and results in the crystal structure being more equiaxed and dense which reduces the number of pores and improves the barrier performance. The danger of this type of coating modification is that the higher density coating has less tolerance to flexing and may have higher intrinsic stress than

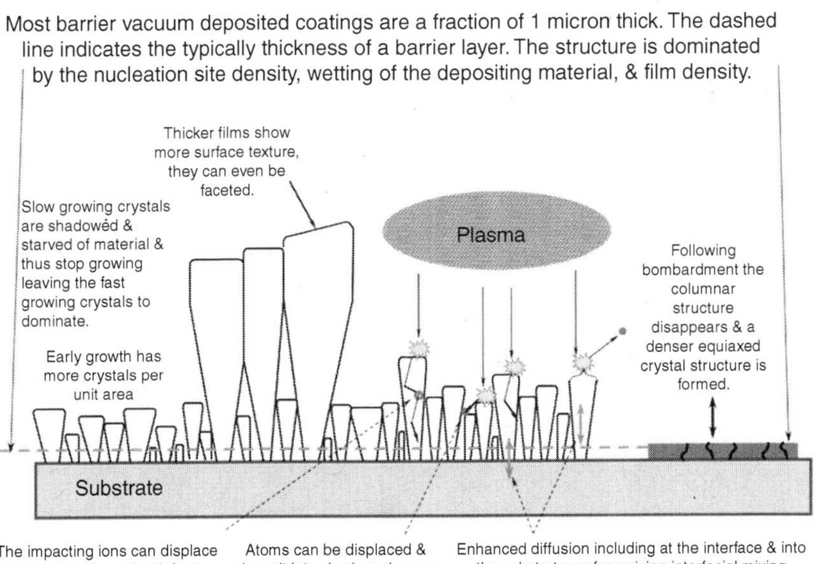

Figure 6.13 A schematic of coating growth & the changes with thickness and plasma modification.

the more porous columnar growth. It is typical to find densified coatings causing problems in winding polymer films as the stress in the coating has caused the polymer film to curl. The polymer film temperature will rise with the increased bombardment from the plasma and the differential shrinkage between the polymer and the vacuum deposited coating can show up in the film curling (19–29). If this differential stress is too high this can also lead to not just curl but buckling and adhesion failure (30). This can be unacceptable for downstream converting processes such as coating and lamination.

As the nucleation is more likely to start at surface defects it is often seen that these same surface defects can be mimicked in the surface of the coatings. Nodules in the top surface can relate back to a defect such as a filler particle protruding from the surface of the polymer film (31–34).

The energy available at the surface can be from a variety of sources. The substrate can be heated directly or can be heated by radiant heat from the evaporation source or by bombardment from excited species such as from a sputtering source or from the vapour passing through an electron beam or additional plasma source. There is a natural difference in energy from each different type of source. Evaporation atoms have no excited species associated with them but have the additional heat radiated from the resistance heated evaporation sources. This compares with the sputtering sources that run at much lower temperatures but have a significant proportion of excited species that means that the depositing material is bombarded so that only well bonded atoms will remain on the surface. This means that the speed of growth is slower but the adhesion is higher and the density of the coating greater. Within the sputtering process the design of the magnetron sputtering source can change the bombardment significantly. If the magnetron source is balanced the substrate heating is minimised whereas if the source is unbalanced the substrate bombardment causing substrate heating is considerably higher. These different sources of energy can affect the nucleation process (35–38) but has to be controlled, usually limited, to prevent substrate damage.

The effect of temperature was reported (39) and is shown schematically on Fig. 6.14 showing that with increasing temperature there is increased atom mobility and this mobility allows the atoms to move more easily to increase the density of the coating and fill in the gaps making the coating more equiaxed and less columnar. This crystal structure can also be affected by changing the deposition pressure. As the pressure increases there can be more gas collisions

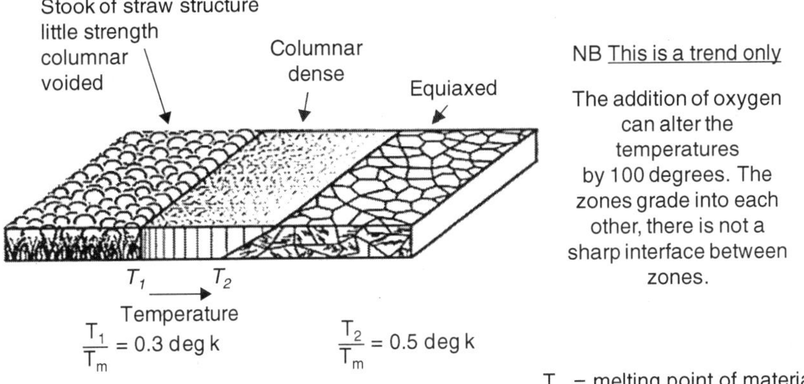

Movchan & Demchishin Phys. Met. Metallography 28, 83 (1969)
This work was for metal films deposited by evaporation to a thickness of several mm.

Figure 6.14 The basic structures produced at different temperatures.

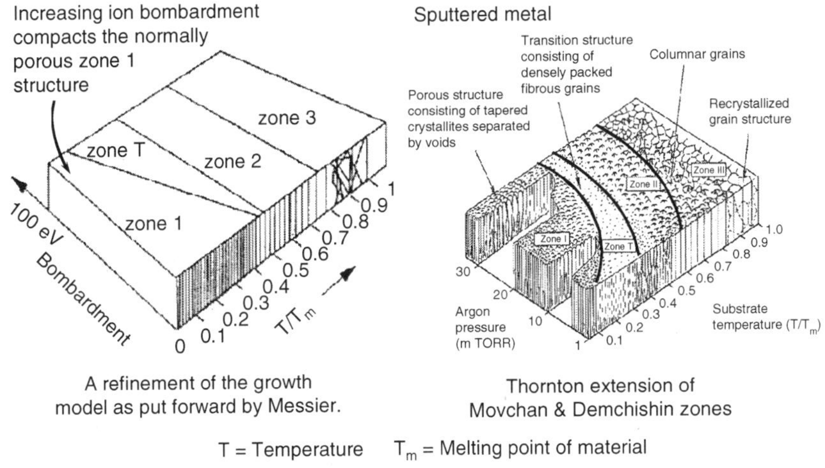

Figure 6.15 The growth modification by pressure and ion bombardment.

that reduce the deposition energy and so the position of the different zones is modified. This variation was plotted (40) and a further extension to the crystal structure zone model by using an additional plasma was also developed (41) this is shown in Fig. 6.15. These zone models allow predictions of what type of structure to expect and what will happen if the temperature or excited species proportion is changed. It is worth reiterating that this structure is only seen in its full extent if the coatings are grown to be very thick. With thin coatings these structures are less easy to determine as the there are too few crystals to make differences between columns and equiaxed crystals easily identified.

Realistically for the vacuum deposition of coatings onto polymer films there is not a wide range of variables we can use. Most of our time and efforts are spent in limiting the effects of heating during the deposition process. With the resistance heated boat evaporation of aluminium the heat load, from the combination of radiant heat from the boats and latent heat of condensation from the depositing aluminium, is enough to take the polymer film temperature up in excess of eighty degrees Centigrade. To limit this rise in temperature the polymer film is usually wrapped around a cooled deposition drum that can be run at sub-zero temperatures which will limit the maximum temperature the film will reach. With evaporation the pressure of deposition has to be kept as low as possible to limit any gas scattering. The material deposition efficiency is low and gas scattering would make it even worse. Increasing the pressure would also make the coating less dense as more gas would be incorporated into the growing coating and the coating roughness would also worsen. So increasing the pressure is not an option. Using an additional plasma has been used but this too adds energy and so either the deposition rate has to be reduced accordingly or additional cooling provided to balance this increase in energy. In most systems the deposition rate is at or near the maximum and so additional cooling is not possible and so this option is also unavailable. Thus most of the options for varying the energy during deposition are not available without compromising the deposition rate and so increasing the coating costs.

This leaves one factor that can significantly influence the nucleation and growth which is the surface energy of the polymer film.

There is also one geometric factor that can have an effect on the nucleation and growth of the vacuum deposited coating. The source-to-substrate distance, the diameter of the deposition drum and the cut-off shielding all combine to have an effect on the coating quality. In Fig. 6.16 the evaporating vapour cloud will start coating the polymer film at point 'A' and finish coating at point 'B'. If the deposition cut off shields are set too wide such that the deposition starts at a near tangent to the deposition drum and grazing incidence onto the polymer surface the adhesion will be poor and the number of nucleation sites reduced. Any surface defects or even surface roughness peaks will shadow the surface and so only allow limited nucleation.

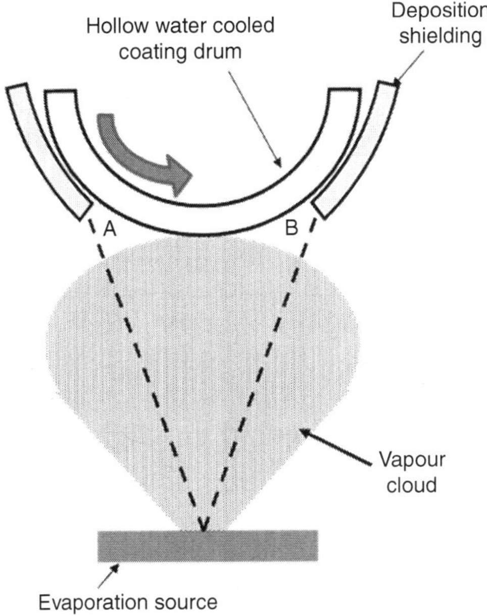

Figure 6.16 A schematic of an aluminium metallizer.

As the film passes around the deposition drum the arrival of the vapour will progress towards normal and once past this mid-point the deposition angle will recede until point 'B' is reached. Again if the deposition cut-off shield is set too wide allowing grazing incident deposition the final layers deposited will be powdery, poorly adhered and will reduce the coating reflectivity or increase haze. As the cut-off shields are usually set symmetrically about the centreline it means that if the shields are set too wide the coating will be both poorly adhered and with poor surface and optical quality (42). Narrowing the width of the deposition shields does reduce the deposition rate and material use efficiency but the quality of the coating improves and the coating porosity is reduced.

Below are some quick guidelines for deposition that summarise the key process conditions.

Increasing temperature leads to increasing crystal size.

This is due to increased atom mobility during nucleation leading to fewer nuclei.

Thicker films will have a greater grain size than thinner films.

Increasing energy leads to increasing temperature leads to increasing crystal size.

Increasing deposition rate leads to decreasing crystal size.
This is due to the mobility of atoms on the substrate during nucleation being limited by more atoms arriving & pinning atoms in place, thus increasing the number of nuclei.

Increasing energy of deposition leads to increasing adhesion.

If the coating is heated to a sufficient temperature it will be annealed.
It will be seen that with annealing the grain size of the coating will increase by recrystallization.

6.2 Managing Heat Load

Polymer films are susceptible to changes when they are heated too much. Some of these changes are permanent such as shrinkage and some are reversible such as tensile strength. These changes often are the limiting factor in the deposition process and the process speed determined by the ability of the system to manage the heat produced in the vacuum deposition process.

Heating of the polymer film is unavoidable and so it has to be managed to control the temperature rise. The exact amount of heat in the process varies with the type of deposition source that is used. One source of heat is present on all vacuum deposition processing and that is the latent heat of condensation which is the heat liberated as the atom that arrives on the surface of the substrate, condenses, stops moving, nucleates and cools. With a resistance heated evaporation boat there is the other significant source of heat that of the radiant heat from the boats (43). For aluminium deposition the boats are usually running at around 1500°C. These boats will radiate heat from all surfaces and so will also heat up the surrounding system and shields. Some of these items may be cooled but others may not. Over the lifetime of the deposition run the temperature of these items may be increasing and they too will radiate heat which may also be seen by the substrate. Thus the temperature seen by the substrate will not be a constant but will rise over the time of deposition. The resistive heated boats do not generate any excited species and so there is no substrate heating from any excited species bombardment. This is also true for induction heated sources too.

Electron beam deposition is similar to resistive deposition in that there is the same latent heat of condensation, a hot molten pool of material in a crucible that will be the source of a radiant heat load but in this case there will be a source of excited species heat load too. The electron beam that is bent to heat the material source hits the material source from the top and some or all of the evaporating material has to pass through this electron beam. In passing through the beam some of the material is ionised and this provides an energetic source of material (44). This slightly increases the adhesion of electron beam deposited coatings over simple evaporated coating because the impact of the energetic material can help remove any poorly adhered atoms or displace atoms to fill voids and so densify the coating slightly. However the penalty is that these energetic species also provide some additional heat.

Magnetron sputtering sources are different as they are not usually running with a molten target. There is still the latent heat of condensation of the depositing material but there is a significantly lower heat load from radiant heating. The magnetron front surface will still be hot as there will be a thermal gradient between the front surface and the cooled back surface but the front surface will be substantially lower than any evaporation source. There will be some stray deposition that will heat up any shields and so if these are not cooled they will heat up and will be a radiant heat source. If the shields are cooled they will contribute little to the substrate heating and the process will be more constant from start to finish of the deposition run. The big difference in the magnetron sputtering process is the substrate bombardment heat load (45–47). The material sputtered from the sputtering target will have most of the energy passed on from the ions that ejected them and this energy is dissipated as the atoms do not simply condense on the substrate but impact on arrival. This impact energy has to be absorbed which generates heat. In addition there may also be heating from some of the ions and electrons from the plasma that is doing the sputtering process. The amount of substrate bombardment by the plasma is dependent on the source to substrate distance and more particularly the magnetic field design. Unbalanced magnetrons are specifically designed to encourage substrate bombardment to heat up the substrate to help in the deposition process. Balanced magnetrons are aimed at minimising the heat load from the plasma and so is the most common type of sputtering source used with the deposition of barrier coatings onto polymer substrates.

There is one other potential source of substrate heating which is from any chemical reaction of the depositing material. Aluminium when it converts to aluminium oxide is an exothermic reaction meaning that it liberates heat. In depositing aluminium opaque metal barrier coatings the typical amount of oxidation of aluminium in the coating is only of the order of a couple of percent and some of this will occur outside the deposition zone and so may be ignored. Where the whole of the aluminium is converted into the transparent aluminium oxide to make a transparent barrier layer this source of heat may also need to be considered.

The most common method of managing the heat load seen by the polymer webs is to wrap the polymer film around a cooled deposition drum so that as the film passes through the deposition zone it is being cooled from the back surface. This is shown schematically in Fig. 6.17. The water can be cooled down to near zero or by adding ethylene glycol (antifreeze) the temperature can be reduced to sub-zero temperatures. It is worth noting that where any cooling is taken down in temperature there needs to be the facility to heat up the surfaces before the system is brought back to atmospheric pressure. If this is not done there will be significant condensation of moisture out of the atmosphere as the surfaces will be below the dew point.

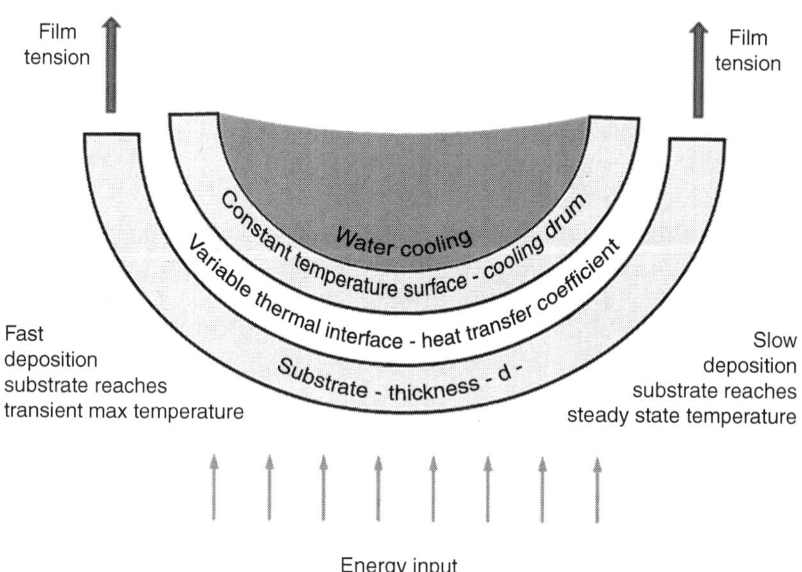

Figure 6.17 The use of a cooled deposition drum for limiting the peak temperature rise of polymer film substrates.

This surface moisture will then make the next pump down much slower. After stopping the deposition if the cooling is then stopped and the temperature is set to above the dew point this will keep the surfaces dry of condensation and help minimise the pump down times. This does require some knowledge of the ambient humidity which will change with the seasons and so what may be a suitable temperature in one season may not be right for another season.

With any deposition process there will be a temperature rise that is dependent upon all the factors specific to the process, materials and system. This will include the substrate material type, substrate thickness and back surface properties, the deposition drum surface roughness and temperature, the heat transfer coefficient and winding tension. For most thin polymer films that are vacuum coated to produce barrier films the residence time through the deposition zone is less than 1 second and so the film never reaches steady state. If the process is slowed down the residence time will increase and if it exceeds approximately 1 second the film is expected to reach a steady state temperature.

To evaluate the heat transfer coefficient it is first necessary to estimate the heat load onto the front surface of the polymer film. This is done as follows.

$$\text{The total specific heat load on a web/foil} = \frac{Q_{total}}{A}$$

where A = coating area

$$\frac{Q_{total}}{A} = \frac{Q_c}{A} + \frac{Q_r}{A} + \frac{Q_p}{A} + \frac{Q_o}{A}$$

where
Q_c = condensation heat load
Q_r = radiation heat load
Q_p = charged particle heat load
Q_o = chemical reaction heat load

The condensation heat load (Q_c)

$$\frac{Q_c}{A} = \gamma \times \frac{t \times v}{l}$$

where			
	γ	= condensation parameter	J/m^3
	t	= coating thickness	m
	v	= winding speed	m/s
	l	= coating length	m

The condensation parameter γ is dependent upon the specific enthalpy h_v of its vapour and the solid density of the deposit ρ

$$\gamma = h_v \times \rho$$

Thus for the deposition of 25 nm of Aluminium at 1200 m/min (20m/s) with a coating zone of 0.5m the condensation heat load would be;

$$\frac{Q_c}{A} = \gamma \times \frac{t \times v}{l} = 3.8 \times 10^{10} \times \frac{25 \times 10^{-9} \times 20}{0.5} = 38\,\text{kW/m}^2$$

Where γ for Aluminium = 3.8×10^{10}

The radiation heat load (Q_r)

$$\frac{Q_r}{A} = \phi\, \varepsilon_s \varepsilon_d \sigma (T_S^4 - T_F^4)$$

where			
	ε_s	=	Emissivity of source
	ε_d	=	Emissivity of deposited film
	ϕ	=	The view factor from the film to the heat source
	σ	=	Stephan-Boltzman constant = 5.67×10^{-8} $W.m^{-2}.K^{-4}$

The emissivity of the deposited film changes during deposition from approximately 0.9, the emissivity of the polymer film down to an emissivity of 0.04 for metals such as Al, Cu or Ag.

Typically for a resistance heated source, running at a temperature of 1400°C and with a source to substrate distance of 0.2 m, one would expect the radiant heat load to be approximately 3 kw/m^2. This is approximately 10% of the condensation heat load.

If we look at Fig. 6.17 we see we have the heating on one side of the polymer web and the heat extraction from the opposite side with the steady state temperature difference between the film surface and the cooling drum defined by the web thickness, the thermal conductivity and the heat transfer coefficient between the web and the drum.

$$T_{wf} - T_d = \left(\frac{d_f}{\lambda_f} + \frac{1}{\alpha}\right)\frac{Q}{A}$$

where T_{wf} = Temperature of front surface of web
T_d = Temperature of the drum
d_f = Web thickness
λ_f = Thermal conductivity of web
α = Heat transfer coefficient

Using the above equation it will be seen that the critical parameter is the heat transfer coefficient. The heat is dissipated into the polymer film and the deposition drum but as the thermal capacity of the thin film polymer is low the bulk of the heat has to be removed by the deposition drum.

As the residence time is short and the film never reaches the steady state temperature the lower temperature that the polymer film will peak at can be calculated as follows.

$$\delta T = \frac{Q_{total}}{A.\alpha.s}\left(1 - \exp\frac{-(\alpha.s)}{(d_f.\rho_f.\lambda_f)}\right)$$

where δT = The difference in web temperature from entering the deposition zone to leaving it.
s = the residence time of the web in the deposition zone (l/v)
ρ_f = polymer web density

This calculation gives us an approximate maximum temperature and we can plot the temperature changes as the polymer film progresses through the deposition process. An example of how this might look is shown schematically in Fig. 6.18 which then shows how using a coolant that can be cooled to sub-zero temperatures can be beneficial. In Fig. 6.18 if we follow the progress of the film we can see that it starts at ambient temperature and is pre-cooled down to the temperature of the cooling fluid. Then as the film enters the deposition zone it is heated and reaches some peak temperature following which the temperature reduces all the time the film remains in

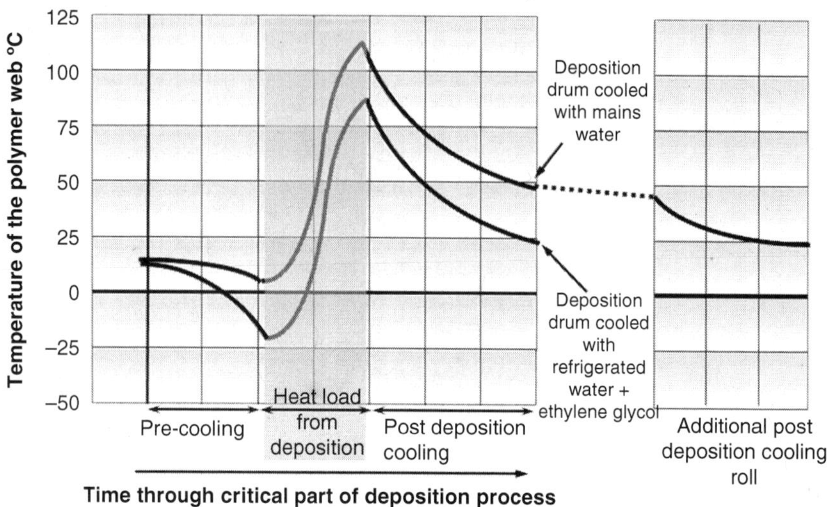

Figure 6.18 The temperature profile of a polymer film through a deposition process whilst wrapped around a cooled deposition drum.

contact with the cooled deposition drum. The peak temperature can be to some extent controlled by the temperature of the cooled deposition drum as shown by reducing the coolant temperature from that of mains water temperature to a sub-zero temperature.

The other thing to note from Fig. 6.18 is the possible need for some additional post deposition cooling. If the polymer film leaves the deposition drum at an elevated temperature the rewind roll will be at an elevated temperature and this can lead to blocking of the roll. The elevated temperature can increase the migration of additives or oligomers to the exposed film surface and speed up the transfer of this material. As the film is wound hot it will shrink as it cools increasing the compression on the core. The film blocking can result in many problems from a higher level of pick-off, film sticking leading to erratic tension or even enough sticking to prevent the roll being unwound. The aim is to have the finished roll wound up at a temperature close to ambient. If the rolls are always coming off hot it may be that one of the rolls after the deposition drum will also need to be cooled to assist in removing the residual heat. All of the post deposition rolls will have a thermal mass and will be helping to bring the polymer film back to ambient temperature but over time these will be being heated by the hot film and their cooling effect will be progressively reduced As shown in

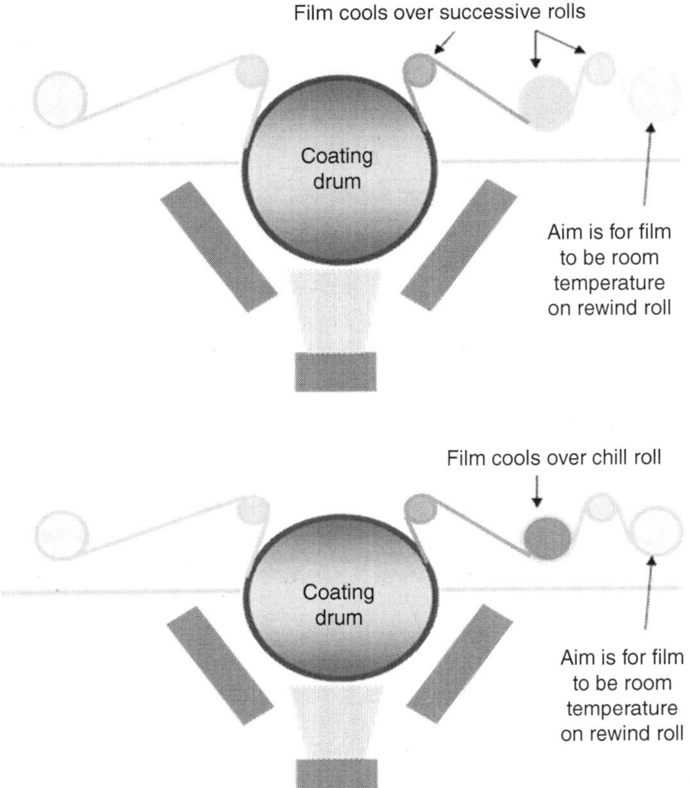

Figure 6.19 A schematic of the use of a post deposition chill roll.

Fig. 6.19. The addition of a cooled post deposition roll will help remove this variation and make sure the roll is wound up close to ambient temperature every time (48).

As polymer webs have a limited heat capacity before the heat relaxes the bonds that result in the polymer shrinking (48) it is common to aim to balance the heat load with at least an equal amount of cooling (49). Although there may be sufficient heat capacity in the cooling system the polymer film will still heat up as the substrate cooling is limited by the ability to transfer heat from the polymer film to the deposition drum, a factor known as the heat transfer coefficient.

The management of heat into the substrate has been the subject of interest over many years (43-57). One of the more interesting findings was the relationship between the surface roughness of the deposition drum and back surface of the polymer film and any trapped gas

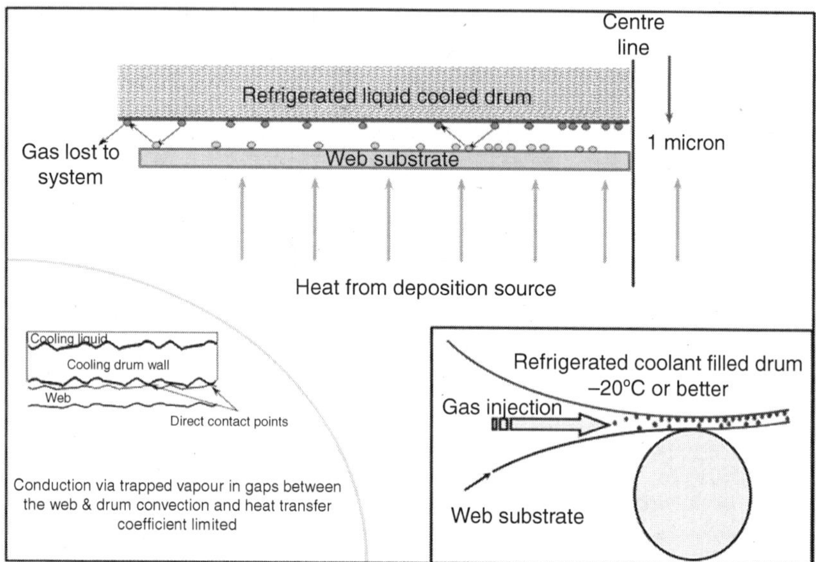

Figure 6.20 The use of gas injected between the polymer film and drum to enhance cooling.

that fills the gaps between the two. It was found that any trapped gas could help to increase the heat transfer coefficient as shown in Fig. 6.20. It had been noticed that polymer films with a higher moisture content would handle the deposition heat load better than drier films. The extension of this finding was to deliberately inject gas into the gap and significantly increase the heat transfer coefficient (58–61).

Measuring the heat transfer coefficient is very difficult to do and more often it is either assumed or inferred from the final web temperature. Modelling of the process has been done and this has improved the understanding of when the gas injection works well and when it does not (62). The process does not work well for narrow film widths or for slow winding speeds. The gas also works as a floatation or lubrication layer that enables the web to more easily move over the deposition drum surface which reduces the propensity of the polymer films to wrinkle as will be described in more detail in the next section. The tension applied to the film can also affect the heat transfer coefficient as higher tensions will compress the gas more and increase the heat transfer capacity. The higher tension will also increase the direct physical contact area as so increase the direct conduction component of the heat transfer coefficient.

It is preferable to use this gas injection system wherever possible as this will improve the consistency of the process. If the moisture content of the polymer film is expected to provide the gas to fill the gaps between the deposition drum and polymer film it will be a variable depending on the moisture content of the polymer. This can be dependent on the history of the polymer film including its manufacture and subsequent storage conditions. It can also depend on any pre-processing of the polymer film inside the vacuum system.

It is preferable to try to minimize the water content within vacuum systems in order to produce cleaner aluminium metal coatings or improve the control of reactive deposition processing. With polymer films there is the air and humidity that is entrained in the wound up incoming roll that needs to be removed but also there may be some water content to the polymer film too depending on the film type. Many systems will include a cryo-panel pump (often referred to as a Polycold system after the leading supplier of such systems) in the unwind zone to getter the water released as the polymer film is unwound. Following this the film may see a plasma cleaning process where the surface bombardment will speed up the release of water that is more tightly bound to the polymer surface. If the polymer contains residual water then once the surface is denuded of water there will be a driving force for water to diffuse from the reservoir of water in the bulk of the polymer film to the surface. The speed of this process will be diffusion rate dependent and so substrate heating is an important factor. After plasma treating the surface the film will then move to the cooled deposition drum where the diffusion rate will be slowed until the film reaches the deposition zone at which point the thermal spike will speed up the water diffusion rate. As there is a coating being deposited onto the top surface of the film this will limit the amount of water that will be released from this surface. On the back surface there is no such restriction and water can be released into the gap between the film and drum and it is this water that will increase the heat transfer coefficient as shown schematically on Fig. 6.21. As this can be a variable it means that for some processes the production can be similarly variable with sometimes wrinkles forming and other times not depending on the film water content. Inserting gas between the film and drum reduces or even eliminates this as a variable.

In some processes there are efforts made to outgas the film prior to the deposition process. In some extreme cases this has included winding the film through at a very low speed under a moderate heat

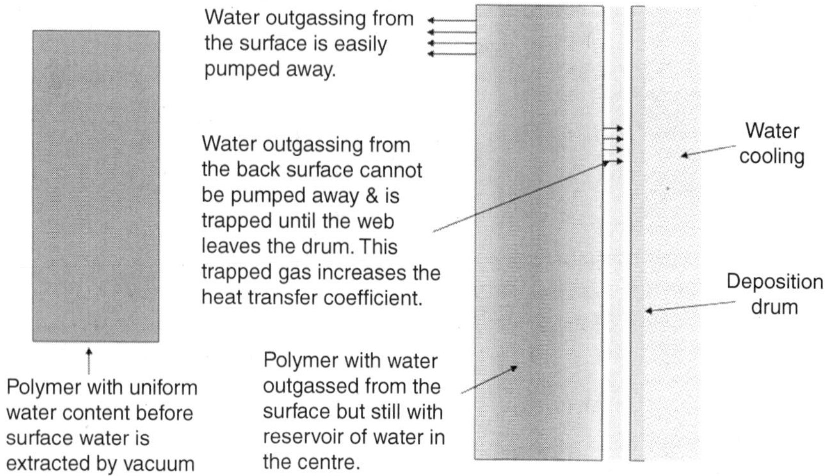

Figure 6.21 Polymer film outgassing due to water content.

load to try to diffuse as much of the water out of the bulk film as possible. This is an expensive choice and needs to be demonstrated as a cost benefit before being included as part of any process. It is easy to demonstrate the effect of water contained in the bulk of a polymer film. If the polymer film is wound quickly through the vacuum system there will be a pressure rise as the winding process is started. If the film is then rewound back at the same speed there will be a similar pressure rise as the first time. This can often be done many times before there is a significant reduction in the pressure rise. What is occurring is that the water from the surface is removed each time but there is sufficient water in the bulk polymer to replace this surface moisture during the time the film is in the roll. Heating the film during each pass and slowing down the winding speed will allow for more water being removed in each pass. However unless the winding speed is slow enough along with heat enough to allow for the water from the centre of the thickness of the film to reach the surface in the time the film travels between the un-wind and re-wind rolls then there will still be some water released the next time the film is un-wound. This effort to minimise the effects of moisture cannot be justified for food packaging films but some effort at outgassing may be justified for some ultra barrier materials and applications.

Heating is not all bad as some of the coating properties can benefit from substrate heating. However for most of the time the polymer substrate is the limiting factor in allowing heat to be used to improve the coating performance. For barrier coatings destined for the food packaging industry where there is a large cost pressure the substrate costs are usually kept as low as possible and so the range of substrate temperature is usually most limited. In ultra barrier type coatings there is an acceptance of higher costs for this premium product and so a higher specification substrate can be chosen. A higher temperature performance polymer may be used such as a polyethylene naphthalate (PEN) film instead of the more typical polyethylene terephthalate (PET) film where the working temperature can be increased. Either the PET or the PEN film could also be heat stabilised (63). This heat stabilising is an additional process that can be done following film manufacturing where the film is taken under low tension to an elevated temperature to allow as much thermal relaxation of the polymer to take place as possible so that on subsequent excursions to the same temperature there will be minimal dimensional changes to the film. This soaking at an elevated temperature takes time and large ovens and so is an expensive option and so cannot be justified for packaging films. Some users have deposited coating onto heat stabilised PEN films at temperatures more than 50 °C above temperatures that are regarded as the normal peak deposition temperatures for PET films that are not heat stabilised.

There is a trend for vacuum coating systems to get bigger and wider and some effects that are minor in smaller systems can become more significant when the dimensions increase. In Fig. 6.22 we see a schematic of a deposition drum. In smaller systems the coolant can be easily fed into the drum by exiting a tube and spraying against the end wall of the drum and returning to the other end and exiting by a concentric return tube. In this case as the coolant flows from one end to the other of the drum it picks up heat and so there may be a small temperature difference from one end of the drum to the other. As the system width increases this temperature difference increases. This is not so important for the deposition of metals but where there is any reactive process this non-uniformity can produce variations in the coating structure, stoichiometry and performance across the width. The same problem can also occur with cooled deposition shields. Often these are not as well cooled as the deposition drum and cooled shields may be linked together in series so that some shields will always be hotter than others. In designing the cooling of any system it is important to make sure the cooling capacity is correct and that the cooling circuits do not lead to thermal gradients that could affect

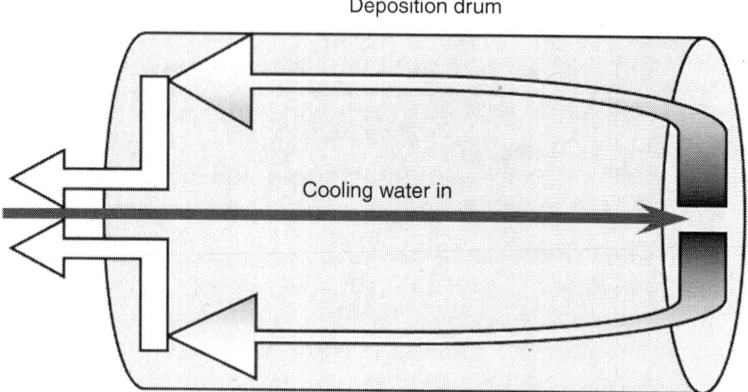

Figure 6.22 A schematic showing how the coolant can vary along the length of the deposition drum.

the deposition process. The cost of cooling can be high and there is always a tendency to have just enough cooling but no spare capacity. Over time the cooling efficiency may drop or the deposition process may be developed to run faster and require more cooling and so the cooling capacity may become limited. In this eventuality there will be a progressive increase in temperature of the system over the course of the deposition run. Again this is not so significant for the metal coatings but can affect any reactive deposition process more with the resultant coating being different at the beginning of the deposition run compared to the end of the run. These longer timescale effects are more difficult to identify and correct once a system is built and running. It is always easier to make sure at the design phase that there is inherent system uniformity and suitable capacity for process development than to correct these at a later stage.

6.3 Web Winding in Vacuum

When winding polymer film at atmospheric pressure there will always be a thin boundary layer of air dragged along with each film surface. This air layer can act as a lubricant as the film passes over rollers and will become trapped between layers when the film is rolled up. These effects disappear when the film is wound in vacuum. The key aspects of winding films in vacuum are the quality of the film being wound, the alignment of the winding system and suitable tension control of the web through the winder.

If we start with the web quality there are variations away from the ideal that can cause winding problems. If the film has been stretched unevenly during biaxial orientation the film will be curved with a short edge and a long edge. Even where the biaxial orientation is even there may still be a residual stress profile (64) which will be symmetric about the film centreline but as the film may be slit from the mill roll into smaller rolls some of this symmetry may be lost as shown in Fig. 6.23.

The tension will first take effect where the web is thickest or where the film is shortest. This can be seen on the winding system by very slowly increasing the tension from zero. As the tension first starts look for where the film becomes rigid first as this will be taking the tension first. If there is a high spot in the profile this could be anywhere across the web and with this area becoming tight the rest of the film will appear slack and may show signs of flutes being present. Fig. 6.24 shows the effect if one edge is shorter than the other. Figure 6.25 shows where the effect is symmetric about the centreline for both long-in-the-middle (LIM) or short-in-the-middle (SIM) films. This symmetric profile may be seen in wide film rolls but after the full size mill rolls have been slit it is common to see one edge a different length to the other as in Fig. 6.24.

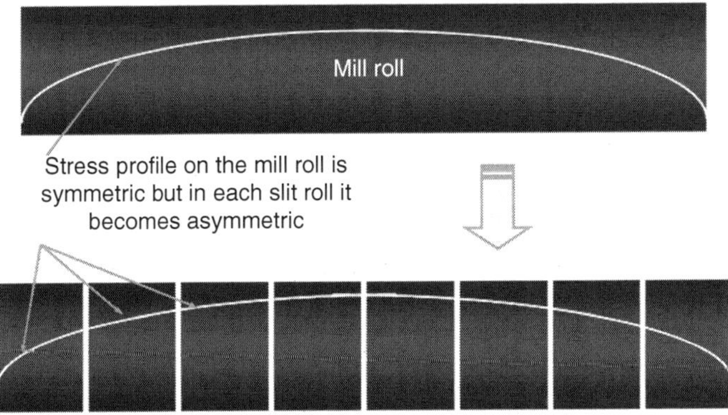

Figure 6.23 A schematic of the residual stress profile in mill rolls and how this will result in differences in slit rolls.

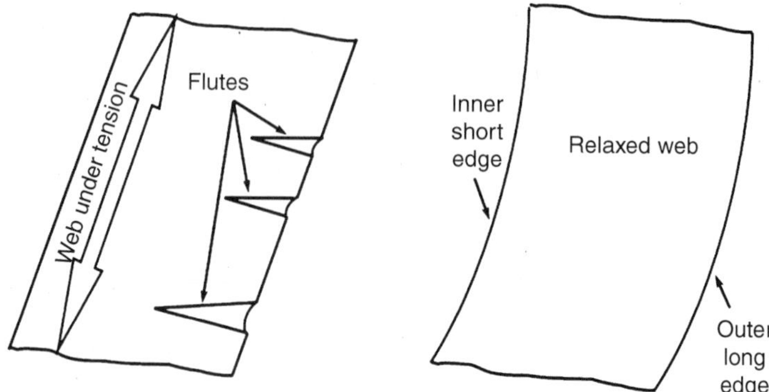

Figure 6.24 Flute in one side of a polymer film due to film curvature.

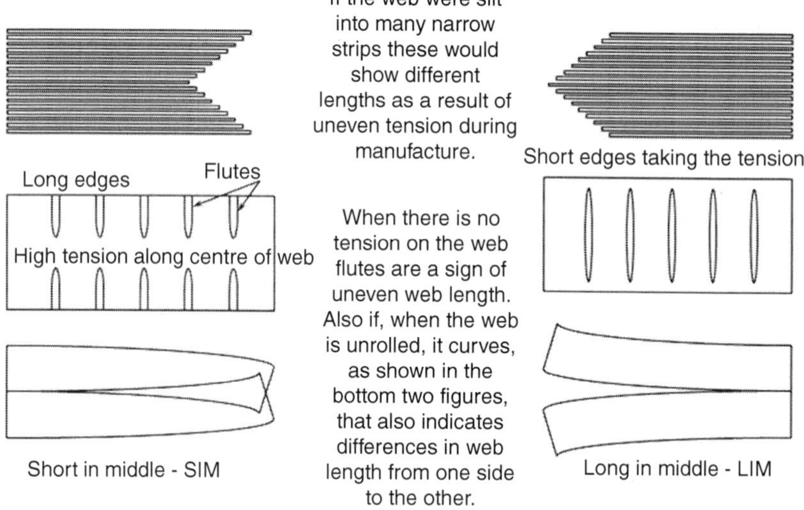

Figure 6.25 A schematic of film length variations.

If there is a profile high spot or if one or both edges has been poorly slit there will be a localised difference in thickness. This high spot in the film profile is known as a gauge band and may or may not be visible when the film is wound in atmospheric pressure but when the air is removed as in winding the polymer film in

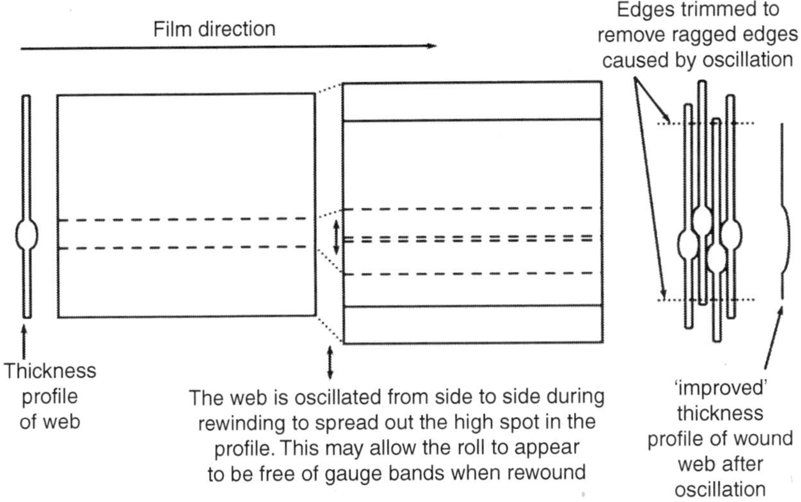

Figure 6.26 A schematic of how film oscillation can help reduce the impact of a film profile high spot. This may hide the problem when the roll is wound in air but the gauge band may appear after the film is wound in vacuum.

vacuum this profile difference can be much more noticeable. Film manufacturers to reduce the visibility of these profile variations may oscillate the film during winding to spread the position of the high spot as shown in Fig. 6.26. A high spot gauge band because it is slightly thicker than the rest of the film when wound will grow the diameter of the roll of film at a faster rate than the adjacent film and all the tension will be concentrated in this gauge band, as shown in the photograph in Fig. 6.27. In some cases this can mean that the gauge band can be tensioned beyond the yield stress and so may have a permanent stretch leaving a pucker along the length of the gauge band.

The hardness of rolls can be quite important. If a roll arrives too loosely wound it can easily telescope on being handled or during the initial vacuum pump down process. If the rolls are wound hard they can highlight any profile variations and so a soft wound roll may immediately raise suspicions of either poor handling or poor quality film. No matter what state of hardness the film arrives in by the time it has been re-wound in the vacuum system it will be a hard roll. This is because there is no air interleaving between the film layers as the film is re-wound. The hardness is in effect a measure of the quantity of air that is wound into the roll.

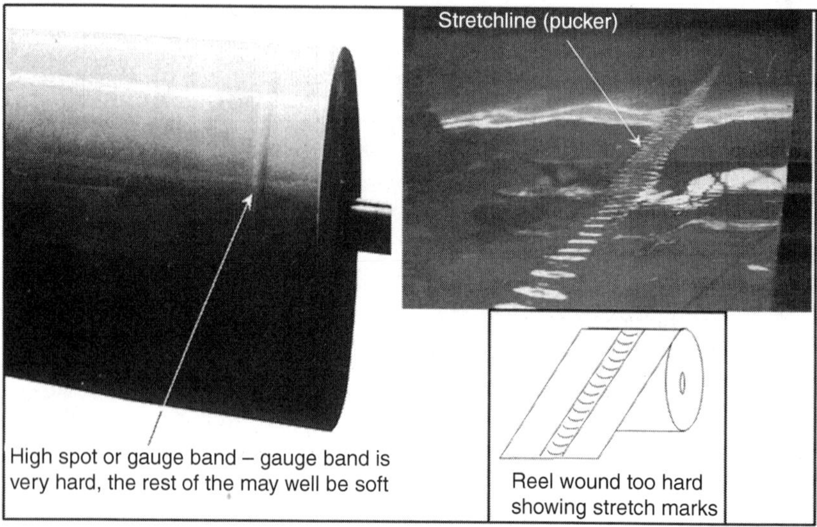

Figure 6.27 A photograph of a gauge band on the left and on the right a pucker from an over tensioned gauge band.

It is common to want to assess the roll hardness and often operators can be seen to knock the rolls with a knuckle to get a feel and listen to the sound as a method of testing the roll. In fact this method has been improved upon by use of a stick (Billy Club) to do the knocking. These methods are subjective and so can be variable from operator to operator and depend on their experience. It is possible to purchase a simple test meter that standardises the same process. There are several products on the market that can be calibrated and give a reproducible measure of roll hardness. The method of testing can vary from having a spring loaded plunger that for a given force will press into the roll and a measure of the depth will give a reading of the hardness. Another test is to use a projectile, such as a ball bearing, that is dropped from a standard height and the height of the bounce is measured. The harder the roll the higher will be the bounce as shown in Fig. 6.28. This same principle can have the projectile mechanically fired so that the test can be carried out at any angle and not just vertically downward.

The following methods or instruments have been described (65–69) the Billy Club and its Variants, RhoMeter and RhoHammer (used to be the Beloit RhoMeter), Backtender's Friend, Schmidt (Concrete) Hammer, Parotester and TAPIO RQP.

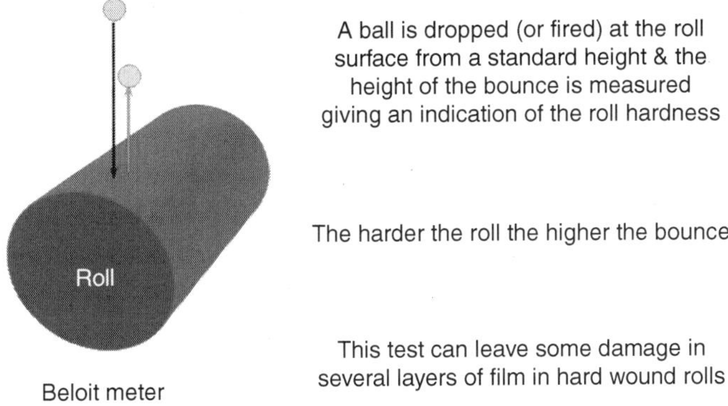

Figure 6.28 A schematic of one of the popular principles of film roll hardness testing.

It is best to measure the roll several times across the whole width so that any variations are picked up. Some of the above meters now have the facility to store many data readings and to be able to transfer the data to computers to make recording the data easy.

One word of caution. Measuring the hardness whether by the knuckle or Billy stick or by automatic machines that bounce a projectile off the surface or press a probe into the surface, all can cause damage to the film. The indentation can penetrate more than 25 layers down into the roll. The harder the roll the greater the depth the damage can penetrate. This damage may only be of consequence to some of the more demanding products such as transparent oxides or similar ceramic coatings where it can be expected to lead to cracking of the brittle ceramic coatings and loss of barrier. Decorative metal coatings may be completely unaffected by this type of testing.

Another source of film thickness variation comes from poor slitting which can raise the edges of the polymer films as blunt knives can damage the edges rather than cleanly cut through the polymer as shown in Fig.6.29. Blunt knives can heat and distort the polymer so that the edge instead of being cut is almost pressure melted and broken through with the hot edge shrinking back and thickening. Blunt knives can cause not only raised edges but also can create polymer hairs or nicks. The hairs are less of a worry than the nicks that can be the source of film tears and polymer film breaks. The slitting is a source of dust that contaminates the polymer film and blunt knives generate more debris than sharp knives and so more pinhole defects can be expected from poorly slit film than from well slit film.

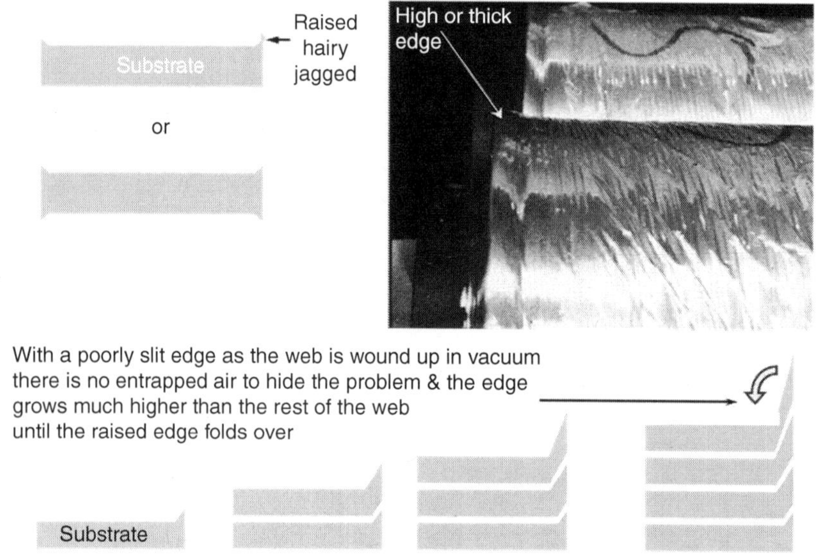

Figure 6.29 A photograph of the effect of poor slitting on the edge of a roll of film and a schematic of how the edge can fold over it the thickening is too great.

It has been my hope for some time that someone will take up the challenge and use laser slitting of the polymer film. Polymer films have been cut to make some metallized flakes and have been pierced to make some breathable films for some food packaging such as baked goods so it is possible to cut the polymer films. I have heard excuses such as being too expensive but nobody has ever been able to prove this to me. Most of the time it is a perception of the cost of lasers which is often based on the cost of lasers some ten or more years ago. The cost and reliability of lasers has improved considerably over the years and I believe that if the full costs are compared I would expect that slitting by lasers would be competitive. When changing the mill roll pattern slit widths lasers can be changed by adjusting the laser direction via mirrors which can be done almost instantaneously. This compares with the re-positioning of knives which can require removing them from a shaft and changing each set of spacers between each knife set. Knives also require maintenance including blade sharpening which lasers do not require. Thus there are a number of cost saving benefits to laser slitting as well as the benefit of reducing the dust generation and so improving the film quality which in turn would convert into improved barrier performance. I would expect that once one supplier breaks ranks and develops cleaner slitting everyone else will have to follow suit to remain competitive.

One of the more common problems that is associated with vacuum deposited coatings is that of polymer film wrinkling (70–80). This is a problem that sometimes gets blamed on the winding system but usually this is false. There are several possible causes of wrinkling but most revolve around the heat load causing a thermal spike in the polymer film. The polymer film when heated will want to expand but it is constrained by having tension being pulled on the film in the machine direction as well as friction between the film and deposition drum.

If we follow the polymer film as it progresses through the deposition process there are various changes in stress that the film experiences as shown schematically on Fig. 6.30. Initially it is common for the web to be placed onto the deposition drum with the help of a spreader roll. The film at this point experiences a tensile transverse stress. The deposition drum is cooled and so the film wants to shrink which, as the film is held in place on the deposition drum by the combination of friction between the drum and film and the machine direction tension. The film when it enters the deposition zone is heated rapidly and the film wants to expand. Initially this expansion overcomes the shrinkage from the initial cooling and then

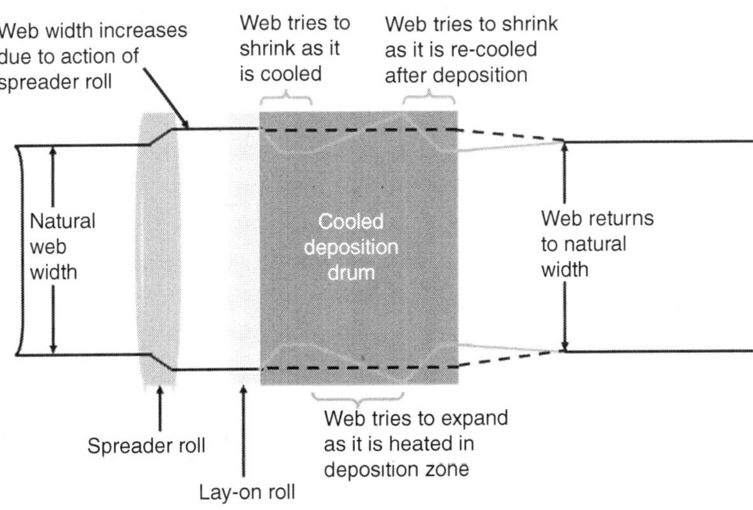

Figure 6.30 A schematic of the changing transverse stress on the polymer film through the deposition process.

it overcomes the tension imposed by the spreader roll and then any further expansion will put the film into transverse compression. This compressive load is where the film is susceptible to buckling and the initiation of wrinkles (also known as tramlines or railroad tracks).

The introduction of gas not only improves the heat transfer coefficient but also reduces the friction between the deposition drum and polymer film. This reduced friction can allow the film to reposition itself on the drum if the transverse tensile or compressive stress becomes too large. This movement when the film is under the compressive stress prevents the film from buckling off the deposition drum and so initiating a wrinkle.

The change in friction is dependant upon the amount of gas injected between the drum and film as shown in Fig. 6.31. The greater the amount of gas injected between the drum and film the greater the reduction in friction. Also the greater the amount of gas injected the higher the heat transfer coefficient. The other parameter that can increase the heat transfer coefficient is to increase the machine direction tension that is pulled onto the film. The greater the tension the greater the compression of the gas which raises the pressure of the gas between the drum and film and this increases the gas collisions and the transfer of heat from the film to the drum.

It is worth noting that debris on the reverse side of the film can also cause wrinkles as the debris prevents the film contacting the cooled deposition drum over the area formed by the 'tent' defined

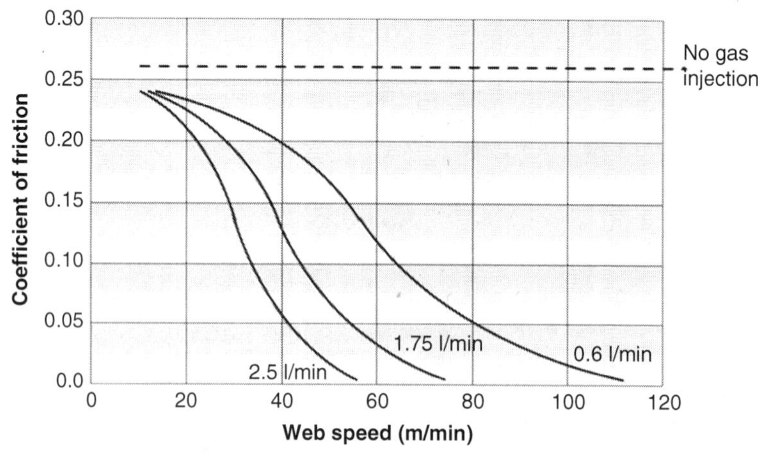

Figure 6.31 The effect of different rates of gas injection on the friction between the deposition drum and polymer film.

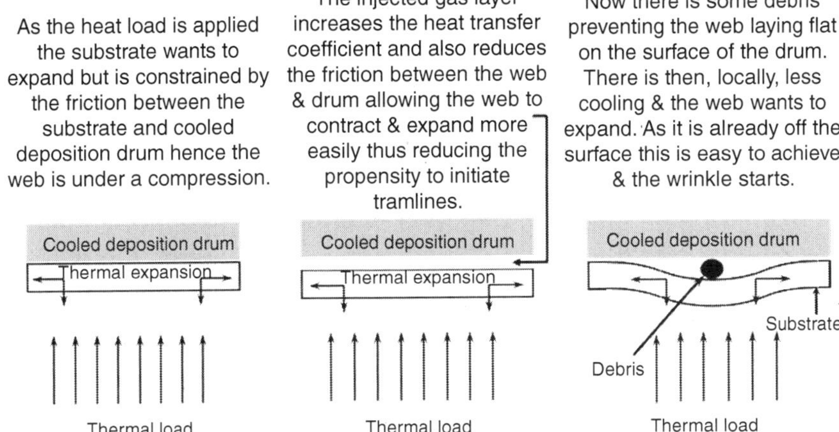

Figure 6.32 The improved cooling and positive effect of reducing the friction between the polymer film and deposition drum and the detrimental effect that debris can have by initiating wrinkles.

by the height of the debris and the tension pulled on the film in the machine direction as shown in Fig. 6.32. As the film is not being cooled over this area the film locally will reach a higher temperature and will expand more than the surrounding film this will increase the compressive stress on the film as well as distorting the film, already starting the film towards buckling behaviour. Buckling once initiated tends to continue and this forms a wrinkle. If the wrinkle stays in the machine direction it is an indication that the film winding is acceptable, however, if the wrinkles are offset from the machine direction then it is indicative of there being a lateral force that could be due to uneven winding. Uneven winding could be caused by one of the rolls not being parallel to the rest of the winding system but more often is an indication of uneven tension due to a profile problem.

The debris on the reverse side of the web can be from the film manufacturing process but can also be from debris settling on the top of the deposition drum during the time the winding system is at atmospheric pressure. Some machines have a temporary shield that protects the top surface of rolls and the deposition drum during the time the winding system is at atmospheric pressure when it is often adjacent to the deposition system that might be undergoing cleaning and hence generating considerable amounts of debris. Debris that either is brought in by the polymer film or is present on the deposition drum may initiate many wrinkles if the debris stays with the deposition drum rather than become attached to the polymer film and hence get taken away to be included in the re-wound roll. Thus keeping the winding system clean is an important part of preventing wrinkling. The use of tacky roll cleaning to make sure that the back surface of the polymer film is clean is beneficial. If the roll of polymer film is going to be cleaned on the front surface it always makes sense to clean the back surface too as debris from the back surface can be transferred onto the front surface after the film is re-wound. If the tacky roll cleaning is going to be done in the vacuum system where the film can be coated before re-winding there may be a temptation to only clean one side but this would leave the film vulnerable to debris induced wrinkling and post deposition coating damage. Thus even in vacuum it still makes more sense to clean both film sides before coating.

Film tension needs to be managed throughout the winding process. As with everything else this can be made easier if the starting point is the supply of high quality rolls. Rolls that are unevenly wound or wound with too little or too much tension will be harder to wind than those wound with the correct tension. Rolls wound too soft can telescope whilst being handled and even once loaded into the vacuum system as shown in Fig. 6.33. Once loaded into the vacuum system and under tension soft wound rolls will want to move as the entrapped air is removed during the pump down. As the film moves any slight variation in film profile will lead to an uneven tension which will put a sideways load onto film and this can lead to the roll telescoping. Once any sideways movement starts the air acts as a lubricant and the telescoping can be exaggerated as shown

Figure 6.33 Some photographs of rolls that have telescoped.

in the right hand photograph in Fig. 6.33. Once this telescoping has occurred there will likely be surface damage which will lead to reduced barrier performance and so even if the telescoping can be corrected by rewinding the quality of any coating is likely to be less than if the roll had not telescoped. It is always worth checking why the roll was wound with low tension. If the film has a poor profile it is more easily hidden if the tension is reduced allowing more air to be entrained into the roll and so keeping the layers further apart allowing for a larger profile variation.

The vacuum system winding system will usually have a minimum of three zones each of which needs to be separated from the others. The three zones will be the un-wind, deposition and re-wind zones. The aim is to control the tension of the polymer film around the deposition drum and to do this well the tension needs to be independent of the influence of the tension in the incoming roll and the tension chosen for the outgoing roll. If we look at the left hand schematic in Fig. 6.34 we can see a simple three drive system where the unwind, deposition drum and rewind are all driven. In this case the tensions cannot be isolated and so any variation in the unwind tension will pass through to the tension around the deposition drum. As the tension of the film around the deposition drum can affect the heat transfer coefficient and film temperature this type of variation makes controlling the deposition process more difficult. By comparison the schematic on the right hand side there is a five drive winding system where not only are the same three items driven but also two wrap rolls are driven too. The wrap rolls are one way of separating the tension applied onto the film either side of the wrap roll. The other, less preferred, method of isolating tensions

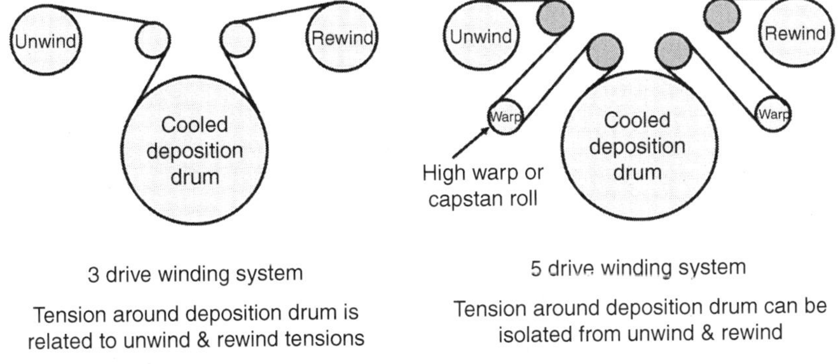

Figure 6.34 Schematics of a simple 3 drive and higher quality 5 drive system.

is to use a nip roll. Nip roll if not well set may end by putting damage into the film. One of the nip rolls needs to be crowned and crowned to match the optimum nip load so that the pressure across the whole width is uniform (81). The two methods of tension isolation are shown in Fig. 6.35. The wrap roll uses the friction between the film and the roll to isolate the tensions on the film either side of the wrap roll. The amount of wrap required around the roll will depend on the friction between the two materials (82). Metal rolls have a lower friction than if elastomer rolls are used. If the polymer film type is changed it is usually possible to change the roll material to change the friction to allow the same amount of wrap to be used rather than needing to change the position of the roll to change the amount of wrap. It is also important to measure the tension in each zone and this is done by using load cells rolls (83). These use simple bi-metallic strain gauges to measure the deflection of rolls caused by the load imposed by the tension pulled on the polymer film. It is important to position the load cells correctly to make sure the load cell measures the maximum deflection of the roll. If the load cell is incorrectly positioned they will be relatively insensitive as the deflection will be smaller than the maximum for any given load. This is shown schematically in Fig. 6.36.

The five drive system with the isolated rewind tension also allows for rewinding with a changing tension as the diameter of the roll increases whilst still keeping the tension around the deposition drum constant. The rewind tension can be varied in many

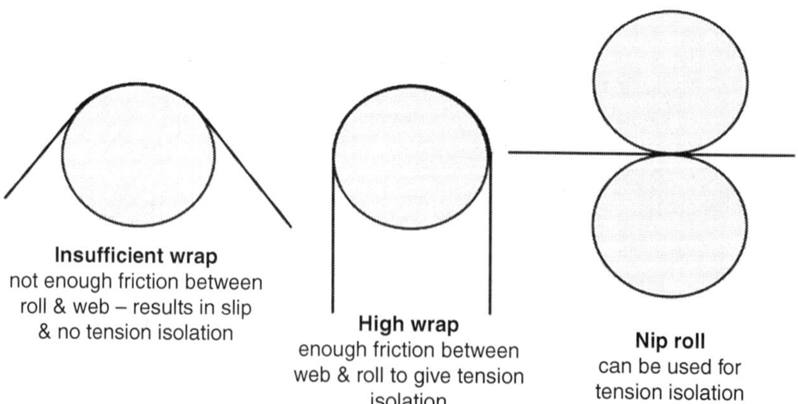

Figure 6.35 A schematic of wrap and nip roll methods of tension isolation.

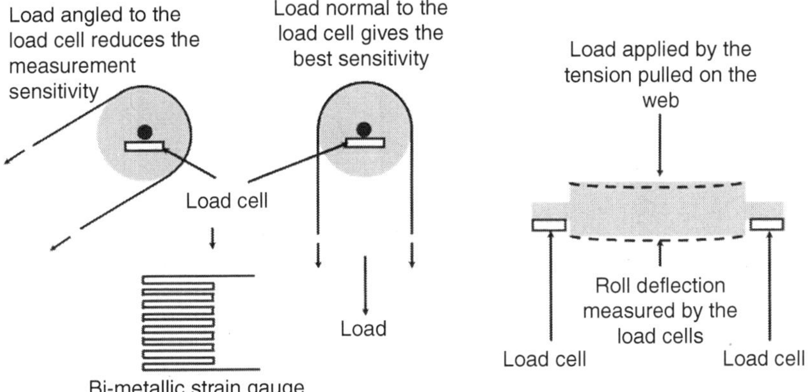

Figure 6.36 The use of strain gauges as load cells to measure tension applied to the polymer film.

ways such as by using a simple constant torque, or by using taper tension, where the tension is reduced with increasing diameter or even using a specific designed tension profile (84). As there is no entrained air the roll will always be wound hard compared to those wound at atmospheric pressure and so it is important to use a low tension otherwise the roll will be very hard and may suffer from problems of blocking.

The polymer film as it passes through the deposition zone is heated and expanded. As the polymer film is wrapped around a deposition drum where there is friction between the drum and film there is some isolation of the tension being pulled at the point where the film leaves the drum and the film across the deposition zone. The film, where it most needs to be held firm against the deposition drum, as it passes through the deposition zone expands and loses pressure and so the heat transfer coefficient can fall and the temperature can further increase. As can be seen in Fig. 6.37 the tensile performance of the polymer film decreases with increasing temperature. Thus if too much tension is pulled the polymer can be stressed beyond the yield point and some permanent elongation caused. If too little tension is pulled the heat transfer coefficient will be too low and the film will reach a higher temperature and have an even lower tensile performance. So as you can see setting the correct tension can be fraught with difficulty. This is made easier for the films that have the gas injected between the deposition drum and polymer film that both increases the heat transfer coefficient as well as providing the

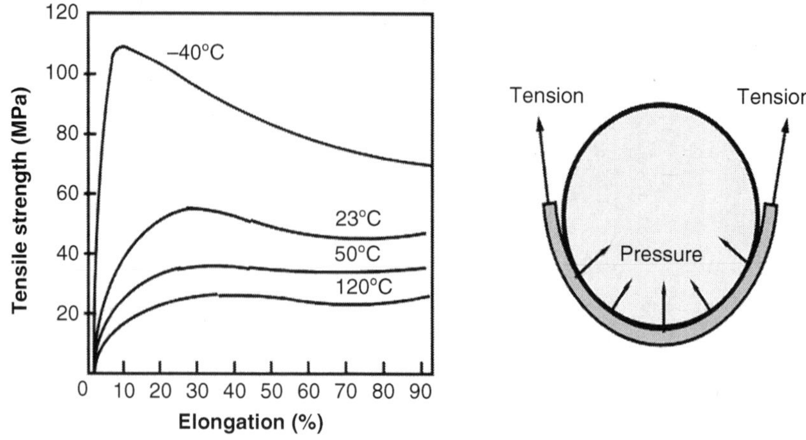

Figure 6.37 The tensile performance with temperature of polyethylene terephthalate.

lubrication that allows the film to float on the drum and so reduces the friction allowing a more constant tension through the deposition zone and hence more constant heat transfer coefficient.

Over the years the winding systems have improved with the improvements in the accuracy of control of the tension through better measurement and control systems. Not only have the control systems been improved but also there is increased precision of measurement of the alignment of the individual rolls available (85–87). The use of a precision gyroscopic measurement system based on the gyroscopic guidance systems used on missiles now makes it much easier to check that all the rolls are aligned and parallel. The system is more precise and easier to use than the engineering slip blocks or laser alignment systems (88). The aim should be to have a roll alignment with a run out of less than 40 microns per metre both axially & in plan view (76,89). As the systems get wider the rolls also need to be designed to withstand bending which can be equivalent to a roll misalignment (90). In vacuum winding systems the distance between rolls is usually short as winding systems are made to be compact to help minimise the volume of the system. As the cost of the pumping system is related to the volume this helps minimise the cost of the pumping system. The polymer films are less likely to wander if the distance between rolls is kept short. If wander is a problem it is possible to add a guidance system (91) although it is always preferable to do without this added complexity. Using crowned or inverse crowned rolls can also be used to keep the film spread and centred on the winding path (92–95). Within the winding system the spreader roll used to minimise wrinkles and position the film onto the deposition drum will also help centre the film (96,97).

Thicker films are stiffer than thinner films and are easier to handle but the trend is towards thinner films and to down-gauging wherever possible. The stiffness increases with a relationship to the cube of the film thickness (98).

Throughout all of these different factors the aim is not to have the film slide over any of the rollers as any relative movement has the potential to put scratches into the surface.

Figure 6.38 shows machine direction (MD) scratches. MD scratches occur when the polymer film is moving at a different speed to the one of the rolls in the winding system. If the scratches are not perfectly aligned in the machine direction but have some angle associated with them then it implies that there is some lateral slip. A common fault that produces MD scratches is where there is a roll that is not driven by a motor but is a tendency driven roll where it relies in the friction between the film and roll for the film to act as

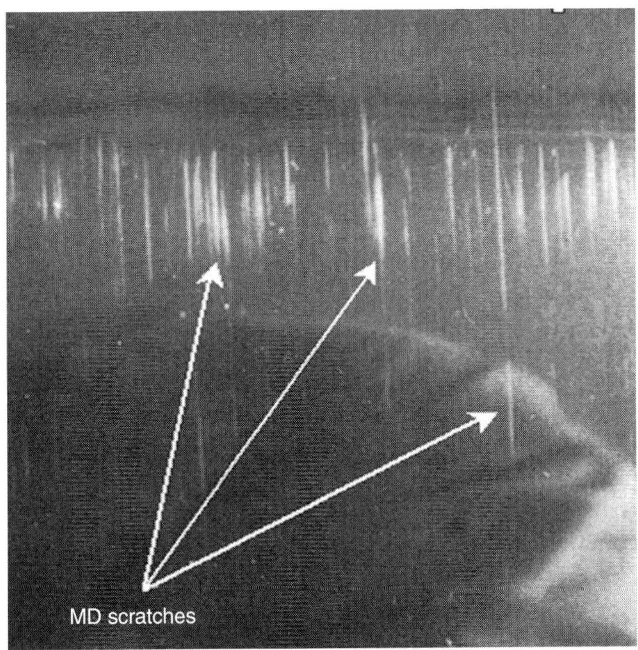

Figure 6.38 A photograph of machine direction scratches.

a drive belt and drive the roll. The time the film is most likely to slip over these rolls is during speed or tension changes.

There is one other incoming material that can adversely affect the winding process and that is the core material that the rolls of film are wound onto. It is critical that these are of high quality as the rolls can never be accurately aligned if they are not machined to tight tolerances or not concentric or circular in cross section. If there is any machining error such as the roll being lobed or tapered it will be impossible to stabilise the tension and the film will always wind badly. The higher the designed winding speeds the more critical the precision needs to be (99). The choice of core material is also influenced by the requirement for coating quality. Spiral wound paper cores are a low cost option but the ends of the core are very easily damaged and shed a large amount of debris that can ultimately end up as pinholes in the coated films (100,101). Slightly better is the lacquered spiral wound paper cores where the lacquer seals the surface of the paper. This is fine until there is any damage to the surface when debris will again begin to shed and become a contaminant. After this there are polymer and metal cores that are a larger cost investment but are cleaner options. Both the polymer or metal cores can still be damaged and this can affect the alignment and winding performance but tends not to be a source of debris. These cores can be made from extruded material but often the extruded material has some curvature and residual stress. The more accurate cores are made from larger section material that is machined down to size. This is more expensive but more accurate. The faster the maximum winding speed and the wider the film the more accurate the cores need to be.

How the cores are located is important. The simple serrated tapered plug is designed to dig into the core to provide grip but this also generates debris particularly from the paper cores. It is better to use one of the other options such as the inflation shaft where a closely fitted shaft, that has expanding segments, can grip the inside of the core without damage.

Handling and storing the cores and rolls is equally important as this too can create variable rolls that will affect the winding. Dropping a metal or plastic core can easily dent the core or make it non-circular. Exposed core ends that can be knocked by fork-lift trucks or other equipment can become damaged and any damage potentially prevents the roll from rotating in a circular path and thus winding uniformly. Protecting the core ends, supporting the roll on the core, rather than resting the roll on the polymer film or on the core ends, will help preserve the quality of the roll and core. Similarly packaging the rolls well with the roll supported on the core and the cores prevented from moving will also preserve the quality of the materials.

Another source of film damage can be from re-using the cores. It is essential to re-use the cores to reduce the costs, particularly for the higher cost machined metal cores. When cutting off the remains of a previous roll it is common to use a knife and there is a temptation to cut against the hard metal core. This is bad practice as the knife will raise a burr and this can damage the next polymer film. The larger the burr the further into the roll the effects can be seen. The film distortion will lead to stress in any coating and the early onset of cracks and loss of barrier. In some cases this will be so near the end of the roll that it may be part of the discarded part of the roll but for some of the ultra barrier type products that are produced at slower rates this material may well be required but will have lower performance than the best material more towards the centre of the roll.

As can be seen from the above the quality of the winding starts with the supply of materials and is dependant on the quality of the materials handling and the hygiene levels employed inside the vacuum system. This attention to detail is essential for the ultra barrier materials but often exactly the same benefits are available for the food packaging products at no extra cost other than improved operator training.

It is important to note that one film that winds badly does not mean you have a poor winding system. If the film runs off to one side a simple check is to turn the same roll over and wind it through again and look at how it winds this time. If the film runs out on the opposite side it indicates the problem is with the film whereas if the run out is still on the same side it could suggest a problem with the winding system. Most of the time winding problems are related to the rolls of film. Once the winding system has been set up it is rare for the rolls to move out of alignment. The changes that can take place are to elastomer rolls where the elastomer can age and harden. This means that elastomer spreader rolls can become less effective over time and may need to be recovered. The elastomer aging can be from heat damage, ultra violet (UV) light damage from the cleaning plasma or simply be because there is a component of the elastomer that is volatile. Any volatile component will progressively evaporate into the vacuum system and the elastomer will dry out. The heat or UV light can damage or crosslink the elastomer and this hardens the elastomer making it less flexible and so the spreading action is reduced.

If there is a concern about the alignment of the winding system it is worth checking if any rolls have been removed and replaced recently. It is rare for winding systems to go out of alignment without any human intervention. So it is also worth checking if there have been any changes to the incoming rolls of material between winding good rolls and poor rolls.

It is easier to wind bad films as it only needs either a poor incoming roll or a poor winding system to produce a poor output roll. Whereas to produce a good roll both a good incoming roll and a good winding system need to be combined to produce good output rolls.

6.4 Troubleshooting

Troubleshooting is a skill that is often overlooked but if not done well can be very costly in both time and money. There appears to be two main causes that lead to struggling to troubleshoot problems. In a number of cases there is the initial difficulty that the problem has been misidentified and everyone has been trying to solve the wrong problem. The second difficulty is the scattergun approach where somebody has an idea what the cause of the problem is and tries a correction. When this fails they then chase the next best idea and so on, following a random walk through possible solutions. Occasionally this can work but often the lack of a systematic approach can be confusing and the true solution can be missed. This problem solving by chasing the next 'best guess' is often compounded by the urgency to get the system back into production.

This has led me to use some simple guidelines to help in troubleshooting which are;

1. Confirm the problem.
2. Question everything and trust nothing. Do NOT make assumptions.
3. Use all senses.
4. Compare to the normal operation cycles or production and look for differences.
5. Ask the question – 'what has been done or changed since the previous successful product'.
6. Be systematic.

7. Do not jump to conclusions, wait for whatever information is available first.
8. Review the process and include learning for future use.

Going through the list it is worth asking the simple or obvious questions just to confirm they have been thought of and checked. This starts with confirming that the problem is real and correctly identified. Do not assume that the questions have been thought of and answered. Others may also have made assumptions and so never bothered to check the reality. An example of this would be if a vacuum system is thought to have no vacuum. This may be true if the gauge is reading correctly but false if the gauge has failed. There is no use in trying to fix the imaginary problem of a loss of vacuum if the gauge is the problem.

Not all of these are applicable in every case such as 'use all senses' which is really designed for the troubleshooting of systems or the deposition process. Often the operators can hear the changing sound of the pumps when different valves operate and can notice when the sound has changed. Pumps may reduce in performance if the drive belts slip and slipping belts can create a burning smell so that a change in smell may be a tell tale sign.

It is only possible to compare production or process cycles if the information has been recorded and stored. Thus this recording and storing of information becomes a critical part of any troubleshooting process. The information recorded should be wider than just the output of any single process. This historic information can include things such as the date and weather, items that at the time look to be of no use at all but may or may not be useful in later troubleshooting. An example of this would be number of pinholes in metallized film that was eventually found to be seasonal due to the higher levels of pollen at certain times of the year which increased the particulate contamination during film production. In vacuum systems the cleanest the system will be is immediately after it is manufactured. Thus the pumpdown is likely to be quickest at this time. This provides the benchmark that future pump down times can be compared to which can highlight the falling performance. Similarly another pump down can be recorded immediately following a system cleaning which again can be used for comparison to confirm that any future cleaning has been done as well. Similar benchmarks can be recorded for other processes. All of this data can be recorded and recalled enabling a comparison to be made for when the product fails to meet the required specification.

This leads to item 4 which is looking for what has changed between producing good product and product that is unacceptable. If the vacuum coating does not meet the specification the deposition may be the first process to be examined but it is worth looking wider to ensure that all the relevant information has been collected and can be considered. This incorporates items 5 & 6 as the manufacturing process is examined. The vacuum coating process does not start and finish with the vacuum system. There is a roll of polymer that enters the vacuum system and the quality of this roll can affect the vacuum coating process. A change in polymer supplier, even for the same nominal grade of film, can result in a change in surface chemistry as well as water and oligomer content. Any pretreatments need to be checked to see if that process is identical or has changed. This can be as simple as a change in season creating a change. Corona treatment is affected by the humidity and when the monsoon season starts there can be a large change in humidity and if the corona treatment is not adjusted the effect of the treatment change can be seen as a change in the adhesion level of the vacuum deposited coating. In this case by trying to solve the problem when only looking at the vacuum deposition process may not have solved the problem. Thus when asking the question 'what has changed?' it should be done in the widest possible way to incorporate the whole process.

To be systematic can be difficult as it may be possible to have several different starting points and then be able to ask questions in many different orders. What is important is not necessarily the order, although this may be helpful, but it is important that all the relevant questions are asked. What can help is every time there is a new problem to troubleshoot is to think through the problem and produce a plan of questions to ask? This can include different sets of supplementary questions that are dependant upon answers to earlier questions.

Once the troubleshooting has been completed this process and list of questions can be refined with the experience gained and this can then become a template for the next time a similar problem occurs. In this way the list of questions can be refined and ordered with regard to importance which will help future troubleshooting. This will also help those less familiar with troubleshooting to learn and gain experience.

Hence it does not matter what the problem is whether it is adhesion, winding, process, uniformity or vacuum the troubleshooting process follows the same pattern as given in the earlier list (102).

References

1. Behrndt K.H. 'Influence of the deposition conditions on growth & structure of evaporated films' *Vacuum* 1963 Vol. 13, pp 337–347.
2. Pashley D.W. et al. 'The growth and structure of gold and silver deposits formed by evaporation inside an electron microscope.' 1964 *Phil. Mag.* 10, p127–158.
3. Campbell D.S. & Stirland D.J. 'The epitaxial growth of silver & gold films by sputtering.' *Phil. Mag.* 1964 Vol. 9, pp 703–707.
4. Moazed K.L & Hirth J.P. 'On the contact angle in heterogeneous nucleation upon a substrate.' *Surface Sci.* 1964 Vol. 3, pp 49–61.
5. Chopra K.L. 'Influence of electric field on the growth of thin metal films' *J. Appl. Phys.* Vol. 37, No. 6, May 1966 pp 2249–2254.
6. Hirth J.P. & Moazed K.L. 'Nucleation processes in deposition onto substrates.' Fundamental Phenomena in Matls Sci. Vol.3 1966 *Surface Phenomena*. pp 63–84.
7. Stirland D.J. 'Electron bombardment induced changes in the growth and epitaxy of evaporated gold films.' *Appl. Phys. Lett.* 1966 Vol. 8, No. 12, pp 326–328.
8. Hill R.M. 'Electrical conduction in discontinuous metal films.' *Contemp. Phys.* 1969 Vol. 10, No. 3, pp 221–240.
9. Chopra K.L. *'Thin film phenomena.'* Pub. McGraw-Hill NY 1969.
10. Morris J.E. 'Effects of charge on the structure of discontinuous thin gold films.' *Metallography* 1972 Vol. 5, pp 41–58.
11. Venables J.A. 'Kinetic studies of nucleation & growth at surfaces.' *Thin Solid Films* 1978 Vol. 50, pp 357–359.
12. Smith R. & Richter A. 'Modelling thin film growth: Monte-Carlo models of fullerite films.' *Thin Solid Films* Vols. 343 & 344 1999 pp 1-4.
13. Siegel R.W. 'Exploring mesoscopia: the bold new world of nanostructures.' *Physics Today* Oct. 1993 pp 64–68.
14. O.S.Heavens *'Optical properties of Thin Solid Films.'* 1991 ISBN 0486669246 Pub. Courier Dover Publications, New York.
15. Stauffer D. 'Scaling theory of percolation clusters.' *Phys. Rep.* 1979 Vol. 54, Issue 1, p. 1–74.
16. Laibowitz R.B. et al. 'Summary Abstract: Cluster formation & the percolation threshold in thin Au films.' *J.Vac.Sci.Tech.* 1983 A. Vol.1, No.2, pp 438–439.
17. Vook R.W. 'Nucleation & growth of thin films.' *Optical engineering* May/June 1984 Vol.23 No. 3, pp 343–348.

18. Freund L.B. & Suresh S. *'Thin film materials – Stress, defect formation & surface evolution.'* 2004 ISBN 0 521 82281 5 Cambridge University Press, UK.
19. Thornton J.A. & Hoffman D.W. 'Internal stresses in amorphous silicon films deposited by cylindrical magnetron sputtering using Ne, Ar, Kr, Xe & Ar+H_2. *J. Vac. Sci Technol.* Vol. 18, No. 2, 1981 pp 203–207.
20. Maniv S. et al. 'Pressure & angle of incidence effects in reactive planar magnetron sputtered ZnO layers.' *J. Vac. Sci. Technol.* Vol. 20, No. 2, 1982 pp 162–170.
21. Fuchs H. & Gleiter H. 'Is the impact velocity of vacuum-deposited atoms of significance for the films structure.' *Thin Solid Films* Vol.101, 1983 pp 55–59.
22. Thornton J.A. 'The microstructure of sputter-deposited coatings.' *J. Vac. Sci. Technol. A,* Vol. 4, No. 6, 1986 pp 3059–3065.
23. Schiller S. et al. 'Processing & instrumation in PVD techniques. *Vakuum-Technik* Vol. 35, No. 4, 1986 pp 35–54.
24. Thornton J.A. & Hoffman D.W. 'Stress-related effects in thin films.' *Thin Solid Films* Vol. 171, 1989 pp 5–31.
25. Neidhardt A. et al. 'Position dependences in planar magnetron sputtering.' *Thin Solid Films* Vol. 173, 1989 pp 109–127.
26. Windischmann H. 'Intrinsic stress in sputtered thin films.' *J. Vac. Sci. Trechnol. A,* Vol. 9, No. 4, 1991 pp 2431–2436.
27. Window B. & Muller K-H. 'Strain, ion bombardment & energetic neutrals in magnetron sputtering.' *Thin Solid Films* Vol. 171, 1989 pp 183–196.
28. Blackwell K.J. et al. 'Investigation of curl .& residual stress in metallized polyimide.' Proc. 5th Ann. Tech Conf. SVC 1992 pp 64–69.
29. Teixeira V. 'Mechanical integrity in PVD coatings due to the presence of residual stresses.' Proc. 3rd Internat. Coatings on Glass 2000 pp 479–489.
30. Weissmantle C. et al. 'Recent developments in ion-activated film preparation.' Ion Plating & Allied Techniques (IPAT) 1979 pp 272–283 Pub. CEP Consultants Ltd.
31. Spalvins T. & Brainard W.A. 'Nodular growth in thick sputtered metallic coatings.' *J. Vac. Sci. Tech.* 1974 Vol. 11, No. 6, pp 1186–1192.
32. Guenther K.H. 'Microstructure of vapor-deposited optical coatings.' *Applied Optics* Vol. 23, No. 21, 1984 pp 3806–3816.
33. Czigany Zs. & Radnoczi G. 'Growth structure & evolution of wavy interface morphology in amorphous multilayer thin films.' *Thin Solid Films* Vols. 343 & 344 1999 pp 5–8.
34. Bochkarev A.A. et al. 'Evolution of crystals during vacuum deposition.' *Thin Solid Films* Vols. 343 & 344 1999 pp 9–12.
35. Robinson C.J. 'The effects of glow discharge treatment on the nucleation of gold on organic substrates.' *Thin Solid Films* Vol.51, 1978 pp L38–L40.

36. Robinson C.J. 'The effects of a glow discharge on the nucleation characteristics of gold on polymer substrates.' *Thin Solid Films* Vol. 57, 1979 pp 285–289.
37. Siewenie J.E. & He L. 'Characterization of thin metal films processed at different temperatures.' *J. Vac. Sci. Tech. Vol.A,* No.4, 1999 pp 1799–1804.
38. Schlemminger W. & Stark D. 'The influence of deposition temperature on the crystalline & electrical properties of thin silver films.' *Thin Solid Films* Vol. 137 1986 pp 49–57.
39. Movchan B.A. & Demchishin A.V. 'Study of the structure & properties of thick vacuum condensates of nickel, titanium, tungsten, aluminium oxide & zirconium dioxide.' *Fiz. Metal.Metalloved.* 28, No. 4 1969 pp 653–660.
40. Messier R. et al. 'Revised structure zone model for thin film structure.' *J. Vac. Sci. Technol. A.* Vol. 2, No. 2, 1984 pp 500–503.
41. Metzner C. et al. 'Emergent technologies for large area PVD coating of metal strips.' Proc. 46th Ann. Tech. Conf. SVC 2003 pp 222–226.
42. Kivaisi R.T. 'Optical properties of obliquely evaporated aluminium films.' *Thin Solid Films* Vol. 97, 1982 pp 153–163.
43. Baxter I.K. 'Effective film temperature control for vacuum web coaters' Proc. 35th Ann Tech Conf. SVC 1992 pp 106–120.
44. Schiller S. el at. 'Progress in high rate electron beam evaporation of oxides for web coating' Proc. 36th Ann. Tech Conf. SVC 1993 pp 278–289.
45. Class W. & Hieronymi R. 'The measurement & source of substrate heat flux encountered with magnetron sputtering.' Solid State Technology December 1992 pp 55–61.
46. Lau S.S. & Mills R.H. 'Temperature rise during film deposition by RF & DC sputtering.' *J. Vac. Sci. & Tech.* Vol 9, No.4, 1972 pp 1196–1202.
47. Thornton J.A. & Lamb J.L. 'Substrate heating rates for planar & cylindrical-post magnetron sputtering sources.' *Thin Solid Films* Vol. 119, 1984 pp 87–95.
48. Baxter I.K. 'Effective film temperature control for vacuum web coaters.' Proc. 35th Ann. Tech. Conf. SVC 1992 pp 106–120.
49. 'Behaviour of Mylar under heat & stress.' DuPont Films information note (1M) 9.94 1994 H-36633.
50. Hendricks W.C. & Diffendaffer P.A. ' Web thermal modelling for vacuum coating processes.' Proc. 38th Ann. Tech. Conf. SVC 1995 pp 134–139.
51. Blackwell J.K. & Knoll A.R. ' Thermal profiles & model for various webs heated in vacuo by IR.' Proc. 32nd Ann. Tech Conf. SVC 1989 pp 91–99.
52. Blackwell J.K. & Knoll A.R. 'Web temperature profiles & thermal resistance measurements of roll sputtered copper & chromium onto polyimide films.' Proc. 34nd Ann. Tech Conf. SVC 1991 pp 169–173.

53. Blackwell J.K. et al. 'DC glow effects on web temperatures' Proc. 33rd Ann. Tech Conf. SVC 1990 pp 194–199.
54. Taylor A. 'Practical solutions to web heating problems' 7th International Conf on Vacuum Web Coating 1993 pp 107–119.
55. Uyama H. et al. 'Thermal behaviour of plastic substrate during deposition.' Proc. 14th Vacuum Web Coating Conf. 2000 pp 63–71.
56. Roehrig M. et al. 'Vacuum heat transfer models for web substrates: Review of theory & experimental heat transfer data' Proc. 43rd Ann. Tech.Conf. SVC 2000 pp 335–341.
57. Affinito J. et al. 'Web substrate heating & thermodynamic calculation method for Li film thickness in a thermal evaporation system. Proc. 44th Ann Tech. Conf. SVC 2001 pp 492–497.
58. Taylor A. & Lievens D. 'Flexible substrates for sputtering.' Proc. 11th Vacuum Web Coating Conf. 1997 pp 238–248.
59. Clow. H. et al. 'Coating apparatus for thin plastic webs' Application Serial No 07/254,088 filed Oct 6th 1988 now abandoned continued as application No 626,320–Patent No. US 5,076,203.
60. Krug T. 'Apparatus & method for cooling films coated in a vacuum.' US patent 5395647 Published March 1995.
61. Casey F. et al. 'Vacuum metallising using a gas cushion and an attractive force' UK Patent GB 2326647 Dec. 1998.
62. Bishop C.A & McCann M.J. 'An investigation of the limitations of the heat transfer mechanism when using a gas wedge between the web & drum.' Proc. 48th Ann. Tech. Conf. SVC 2005 pp 626-630.
63. MacDonald W.A. 'Engineered films for display technologies' *J. Matls. Chem.* 14, pp 4–10, 2004.
64. Tsunanhima K. et al. 'stretching conditions, orientation and physical properties of biaxially oriented film' in '*Film Processing*' Eds. Kanai T & Campbell G Pub. Hanser/Gardner Publications Cincinnati, 1999, pp 339–343.
65. Good J.K.G & Roisum D.R. '*Winding: Machines, mechanics & measurements*' Pub. DEStech Publications Inc. 2007 ISBN: 978-1-932078-69-5.
66. www.testingmachines.com Schmidt Hammer.
67. www.millassist.com/rhometer.htm Rhometer.
68. www.tapiotechnologies.fi/paper_roll_hardness_more.html TAPIO RPQ.
69. www.proceq.com PAROtester.
70. Schwarz W. & Wagner W. 'Thermal limitations in roll coating processes.' Proc. 28th Ann. Tech.Conf. SVC 1985 pp 28–41.
71. Wales J.L.S. & Clow H. 'Model for thermal creasing in roll to roll vacuum metallising.' 2nd Intl Conf. on Vacuum Web Coating. Oct 1988. pp 204–214.

72. Clow H. 'A model for thermal creasing and its application to web handling in roll to roll vacuum coaters' Proc. 32nd Annual Tech Conf. SVC 1989 pp 100–103.
73. Benson R.C. 'Simulation of wrinkling patterns in webs due to non-uniform transport conditions' Proc. 2nd Intnl. Conf. on Web handling OSU 1993.
74. Chiu H.C. et al. 'Mechanical & thermal wrinkling of polymer membranes' ASME Winter Ann. Meeting 1993.
75. 'Behaviour of Mylar under heat & stress.' DuPont Films information note (1M) 9.94 1994 H-36633.
76. McCann. M.J & Jones D.P. 'Web coating dynamic & wrinkling model.' Proc. 41st Ann Tech. Conf. SVC 1998 pp 412–417.
77. Roehrig M. et al. 'Vacuum heat transfer models for web substrates: Review of theory & experimental heat transfer data' Proc. 43rd Ann. Tech.Conf. SVC 2000 pp 335–341.
78. Uyama H. et al. 'Thermal behaviour of plastic substrate during deposition.' Proc. 14th Vacuum Web Coating Conf. 2000 pp 63–71.
79. Hendricks W.C. & Diffendaffer P.A. ' Web thermal modelling for vacuum coating processes.' Proc. 38th Ann. Tech. Conf. SVC 1995 pp 134–139.
80. McCann M.J. et al. 'Buckling or wrinkling of thin webs off a drum' Proc. 47st Ann Tech. Conf. SVC 2004 pp 638–643.
81. Hawkins W.E. 'Vacuum coater thread path design using basic web handling principles & techniques.' Proc. 42nd Ann. Tech. Conf. SVC 1999 pp 430–432.
82. Schwarz W. 'Design aspects of vacuum web coaters–high speed web handling.' Proc. 32nd Ann. Tech. Conf. SVC 1989 pp 156–160.
83. Montalvo III W.W. & Alf C.G. 'Tension & web control in vacuum coaters (metallizers).' Proc. 6th Vacuum Web Coating Conf. 1992 pp 35–44.
84. Taylor K.A. & Pan C.H.T. 'Mechanics of web transport systems for vacuum metallization.' Proc. 29th Ann. Tech. Conf. SVC 1986 pp 136–167.
85. Sasikumar A. 'Measuring roll parallelism in the laminating/coating industry using inertial alignment equipment' Proc. AIMCAL Fall Technical Conference 2007, 21st Internat. Vacuum Web Coating Conf.
86. Tannan J. 'Measuring Roll Parallelism in a Vacuum Metallizing Chamber Using Inertial Alignment Equipment Proc. AIMCAL Fall Technical Conference 2007, 21st Internat. Vacuum Web Coating Conf.
87. Hegde D. 'A return on investment of a roller alignment survey' Proc. AIMCAL Fall Technical Conference 2008, 22nd Internat. Vacuum Web Coating Conf.

88. Walker T.J. 'When rollers fight, webs lose.' *Paper, film & foil converter.* Apl 2004 p 26.
89. Roisum D.R. 'Wrinkling of thin Webs.' Proc. 41st Ann. Tech. Conf. SVC 1998 pp 406–411.
90. Nadolney D.B. 'Taming the wild web: part I.' *Paper, film & foil converter.* March. 1997 p 35–37.
91. Hawkins W.E. 'Controlling your web from beginning to end–part II.' *Paper, film & foil converter.* Jan. 2001 p 52–55.
92. Weiss H.L. 'Various web-control devices minimize wrinkling problems.' *Paper, film & foil converter.* Oct. 1989 p 104–107.
93. Weiss H.L. 'Installing roller devices can prevent web wrinkling.' *Paper, film & foil converter.* Nov. 1989 p 131–134.
94. Weiss H.L. 'Web spreaders help prevent wrinkling in nipping unit.' *Paper, film & foil converter.* March. 1994 p 68–71.
95. Taylor A. 'Practical solutions to web heating problems.' Proc. 7th Vacuum Web Coating Conf. 1993 pp 107–119.
96. Hawkins W.E. 'Under the layon roll: the genesis of MD wrinkles.' *Paper, film & foil converter.* June. 1998 p 32.
97. Hawkins W.E. 'Winder considerations: web spreading at laydown.' *Paper, film & foil converter.* Oct. 1998 p 28.
98. Hawkins W.E. 'Web stability & alignment requirements.' *Paper, film & foil converter.* March. 1998 p 28.
99. Hawkins W.E. 'Winding technology–Part II: the role of the core.' *Paper, film & foil converter.* Sept. 1999 p 32.
100. Kearns R.W. & Munroe J.H. 'Consider film cores when analysing costs' *Paper, Film & Foil Converter* Aug 1995 pp 54–55.
101. 'The need for quality cut cores when winding' *Paper, Film & Foil Converter* June 1998 p38.
102. Bishop C.A. '*Deposition onto webs, films and foils*' Pub. William Andrew Publishing (now Elsevier) 2007 , pp 439–457.

7

Vacuum Deposition

Using an expensive vacuum deposition process rather than an atmospheric coating process has to be for some good reason. Materials like aluminium are very reactive and prone to oxidation and so by depositing the aluminium in a vacuum where the quantity of oxygen or water vapour is limited allows the metal to be deposited without significant oxidation. Other materials deposited for barrier applications may be ceramics that have a very high melting and boiling temperature. In the same way that the boiling point of water is reduced at the top of Mount Everest because of the reduced pressure the same fact can be used to good effect by melting and evaporating metals or compounds at lower temperatures because they are in a vacuum. As the pressure is reduced the number of gas atoms inside the vacuum system is reduced. This means that the distance between gas atoms colliding with another gas atom increases. This distance between gas collisions is known as the mean free path (mfp). With a suitably low pressure (vacuum) any material emanating from a source will reach the substrate without colliding with any gas atoms and so they arrive at the substrate will all the energy they started with. By removing the gas from the system it is also possible to inject some other gas and strike up a plasma and use the energy and gas to modify the chemical composition of the substrate surface or to modify the growth of the coating. This includes converting elements into compounds by reactive deposition processing.

Working in vacuum there is no shortage of vacuum deposition methods. These can be grouped in various ways of which physical vapour deposition (PVD) and chemical vapour deposition (CVD) are the two most prominent groups. Not only can inorganic materials be deposited in vacuum but also organic coatings can be deposited. This then leads to the opportunity of depositing multilayer coatings of both multiple inorganic materials or mixtures of organic and inorganic materials. This wide variety of deposition methods gives rise to a range of barrier coating options from simple single layer deposition at high speed through to multilayer mixed materials in either single or multiple pass deposition processing at slower speeds that can deliver ultra barrier performance.

7.1 Physical Vapour Deposition (PVD)

Physical vapour deposition covers the methods that have, as a starting point, solid materials as the source. The solid material can be heated, causing melting and evaporation, by a variety of means or, as in the case of sputtering, bombarded with energetic gas that can cause atoms to be ejected from the solid source.

7.1.1 Resistance Heated Evaporation

The most widely used method of producing barrier coatings is the resistance heated boat evaporation of aluminium. These systems are referred to as metallizers. The basic operation is that an aluminium wire is fed into a resistance heated evaporation boat where the aluminium melts, forms a molten pool from which aluminium evaporates and the vapour condenses onto the surface of the polymer film that is passing through the vapour cloud as shown in Fig. 7.1. The resistance heated boats are a powder compact material of a mixture of different materials. It is not possible to have a single boat to deposit across the whole width of the polymer film and so a series of boats are used to coat the polymer. Typically the spacing of the boats is around 100mm and so for the largest of the metallizers of 4.45 m width there will be approximately 45 boats each with its own wire feed.

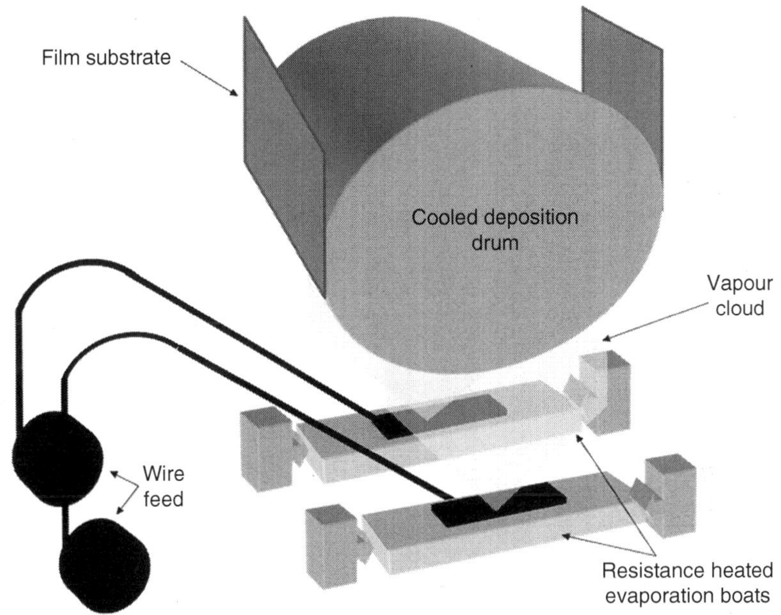

Figure 7.1 A schematic of a wire fed resistance heated boat evaporation process.

If we look at the process and where it can affect barrier coatings we start with the quality of the incoming materials such as the aluminium wire and the ceramic boats and then we can look at how they are used in the process.

Aluminium wire is not pure aluminium but has some other elements included. There are different grades of aluminium that can be used with different levels of purity as per the example shown in Table 7.1. The wire is available in different diameters and with different levels of temper which relates to the wire stiffness (1). Aluminium wire used for evaporation is usually 99% pure or greater. The code used for specifying aluminium is usually 4 digits long. The first digit is generally a '1' and specifies the basic 99% purity, the second digit is used where there is specific control of one particular impurity and the final two digits reflect the two digits to the right of the decimal point of the purity value. Thus 1199 would indicate a 99.99% purity aluminium wire.

The wire can be supplied in different states of hardness. If the wire is full annealed or soft tempered it is graded 'O' and if it is fully hardened by strain hardening to 75% it is graded H18. In between these two extremes are quarter hard H12, half hard H14 & three quarters hard

Table 7.1 Details of typical aluminium wire purity and temper.

<table>
<tr><th>Wire Temper Designation</th><th colspan="4">Description</th></tr>
<tr><td>O</td><td colspan="4">Soft temper - Worked to final dimension then annealed</td></tr>
<tr><td>H12</td><td colspan="4">Strain hardened to 1/4 tensile strength between 'O' & H18</td></tr>
<tr><td>H14</td><td colspan="4">Strain hardened to 1/2 tensile strength between 'O' & H18</td></tr>
<tr><td>H16</td><td colspan="4">Strain hardened to 3/4 tensile strength between 'O' & H18</td></tr>
<tr><td>H18</td><td colspan="4">Hard temper - full hard - strain hardening ~ 75%</td></tr>
<tr><td colspan="5">Alloy 1100 = 1st digit means minimum 99% al
2nd digit refers to any special content control
3rd & 4th digits = % purity of al following decimal point</td></tr>
<tr><th>Alloy/Impurity</th><th>1100</th><th>1350</th><th>1188</th><th>1199</th></tr>
<tr><td>Silicon (Si)</td><td>0.95 Si+Fe</td><td>0.1</td><td>0.06</td><td>0.006</td></tr>
<tr><td>Iron (Fe)</td><td></td><td>0.4</td><td>0.06</td><td>0.006</td></tr>
<tr><td>Copper (Cu)</td><td>0.05–0.20</td><td>0.05</td><td>0.005</td><td>0.006</td></tr>
<tr><td>Manganese (Mn)</td><td>0.05</td><td>0.01</td><td>0.01</td><td>0.002</td></tr>
<tr><td>Magnesium (Mg)</td><td></td><td></td><td>0.01</td><td>0.006</td></tr>
<tr><td>Chromium (Cr)</td><td></td><td>0.01</td><td></td><td></td></tr>
<tr><td>Zinc (Zn)</td><td>0.10</td><td>0.05</td><td>0.02</td><td>0.006</td></tr>
<tr><td>Titanium (Ti)</td><td></td><td></td><td>0.01</td><td>0.002</td></tr>
<tr><td>Others</td><td>0.15</td><td>0.1</td><td>0.01</td><td>0.003</td></tr>
<tr><td>Aluminum (Al)</td><td>99.00</td><td>99.5</td><td>99.88</td><td>99.99</td></tr>
</table>

H16. The temper determines the wire hardness and stiffness which indicates how easily the wire can be bent from the drum of wire round to the boat. As the wire approaches the boat it will be heated and may sag and so some degree of temper may be advantageous.

The wire cost will reflect the processing and so the higher purity takes more processing (2). The wire is produced by drawing a large diameter aluminium rod down to the desired wire diameter. The drawing process work hardens the aluminium and so periodically during the process the wire needs to be annealed with the final annealing being to give a controlled hardness to the final wire.

All aluminium oxidises and the oxide that forms takes up more volume and is under compression and that provides the barrier to oxygen that slows down further oxidation. The oxidation of aluminium is shown in Fig. 7.2. The age of the wire and the storage conditions will affect the thickness of the oxide that is present on the wire surface. The thickness of the wire will also affect the ratio of the oxide to wire with the thinner the wire the greater the proportion of oxide present. Thus a high purity wire that has been manufactured and cleaned recently will help to minimise any spitting during the deposition process. Reducing spitting will help minimise the pinholes in the deposited coating and hence maximise the barrier performance.

To reduce the power and maintain the surface quality lubricants are used during the drawing process. This lubricant is a contaminant to the aluminium evaporation process. It can migrate into the grain boundaries in the wire surface and so the quality of the cleaning of

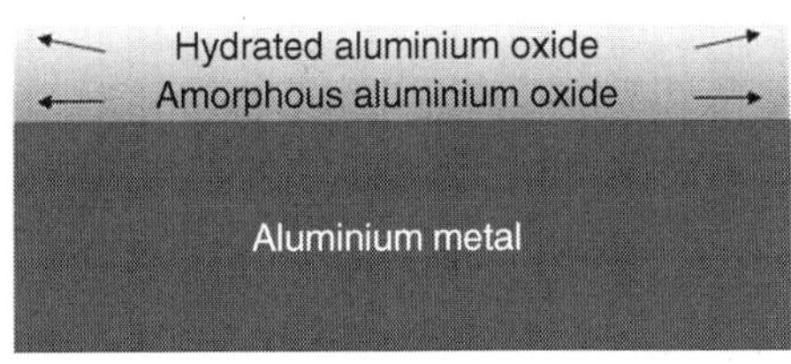

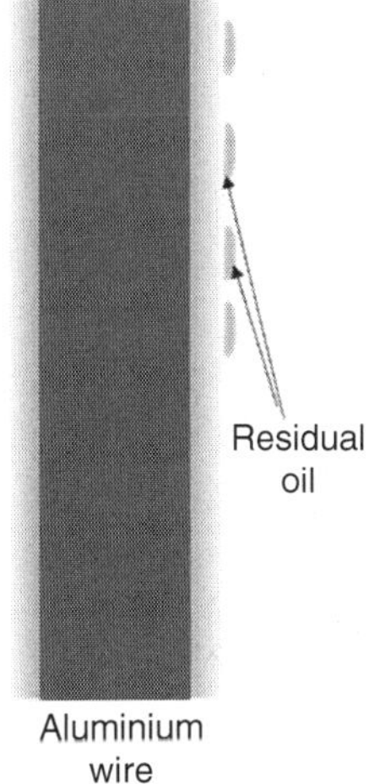

Figure 7.2 The oxidation of aluminium wire and potential contamination from the lubrication during manufacture.

the wire is important and cleaning off the lubricants needs to part of the wire specification.

The minor residual materials in the aluminium composition, the surface aluminium oxide, or hydroxide, and any residual lubricant can all lead to a reduction in the quality of barrier coatings. As the aluminium wire melts and forms a molten pool the contaminants will form a slag or crud on the molten pool surface and tend to collect towards the edges of the pool. If the pool shape or size changes this crud tends to be ejected from the surface, often as incandescent particles as shown schematically in Fig. 7.3. These particles will cause a pinhole in the coating, in fact with some particles there is enough mass that the incandescent particle can burn a hole through the polymer film. This is still referred to as a pinhole but is a full hole through the coating and film substrate and is a weakness in the film that can sometimes cause tears or film breaks in any of the downstream processing.

The quality of how the wire is fed into the hot boat is important too. As the boats are very hot the coils of wire are generally hidden below the evaporation zone and the wire is pulled up and fed through some nip rolls and often guided through a fixed tube so that the unsupported wire travels towards a given point in the hot boat. This can be an area that reflects the care the operators give to setting up the process. The tubular wire guides can become coated with evaporating aluminium and eventually will be blocked and so initially slow down the wire feed and with continued deposition

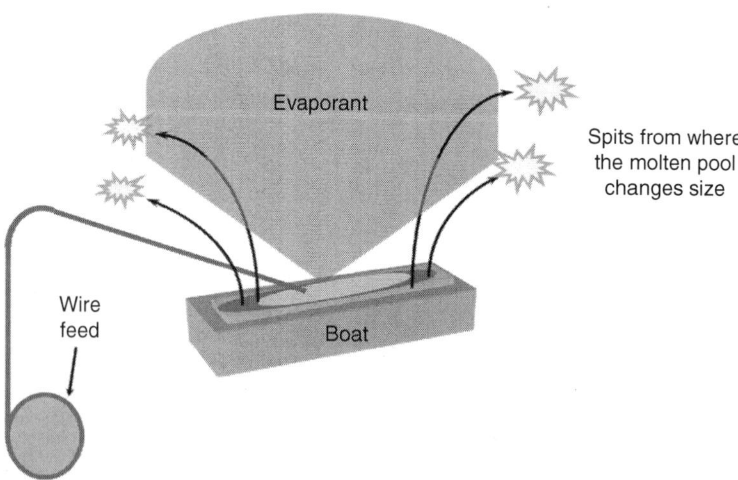

Figure 7.3 A schematic of how molten pool size variations can lead to spitting.

will stop the feed. These tubes need to be regularly cleaned to prevent this problem. The wire thickness and temper will determine how easy it is for the wire to be bent around through the tube. Changing the grade, thickness or temper of the wire will affect how well the wire feed through the tube and how easy it is for the wire to reproducibly reach the boat in the correct position. Once the wire leaves the tube and is in free space it will be being heated by the radiant heat from the boat and this will soften the wire. The unsupported length of wire with the weight of the wire and the increasing softness along the wire will mean that the end of the wire will droop as it closes in on the boat.

The amount of droop will depend not only on the wire diameter as shown in Fig. 7.4 but also on the amount of heat the wire sees from the source. Hence if the source is driven harder at a higher temperature the wire is likely to droop more and more likely to drip from the wire end. A lower temperature source and the wire may well enter the molten pool before melting completely.

Another small problem that has been noticed on some feed systems under certain process conditions is where a knurled nip roll feed system is used. The wire may be soft enough that the knurling

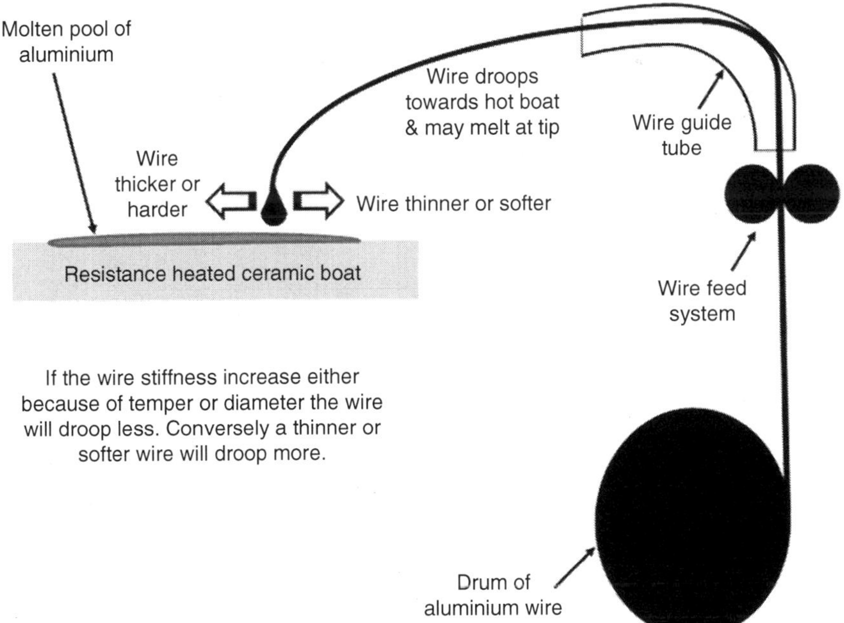

Figure 7.4 The effect of wire stiffness on feed position onto evaporation boat.

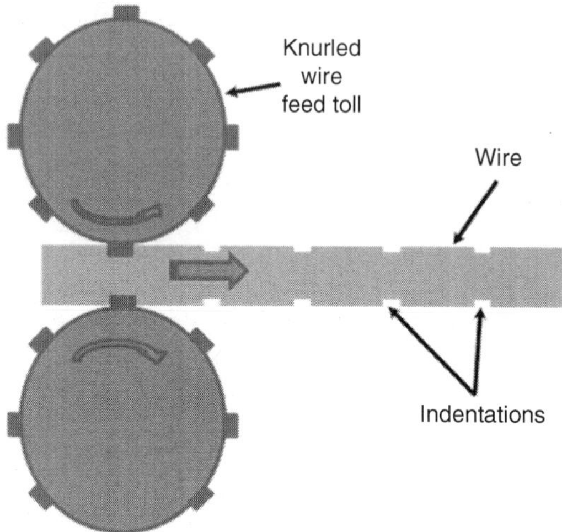

Figure 7.5 A schematic of a knurled nip roll as part of the wire feed.

produces indentations in the wire and next to each indentation would be a slight thickening of the wire from the displaced material from the indentation as shown in Fig. 7.5. Depending on the depth of the indentations this changing wire thickness can affect the molten pool size. If the pool is small then as the thin section of the wire enters the pool the pool size will be a minimum and then as the bulged part of the wire following enters the pool size will grow. This oscillating size will repeatedly move the crud at the pool edge backwards and forwards over the hotter boat surface where flash evaporation can occur and which is where the source spitting originates. Smooth nip rolls may not have as much grip as the knurled variety but do not damage the wire surface. The lack of grip from smooth feed rolls can mean that if the guide tube becomes clogged and the friction increases the wire feed will slow down and become difficult to keep at a constant speed. The variability due to indentations by the knurled roll is not often identified with all the other variations that can occur in the system. If the molten pool volume is increased the additional volume variation between the indented cross section and the bulged cross section may represent a very small change in pool size and result in little or no change in the spitting rate.

Early machines had all the spools on a single shaft which meant that all the wire feeds were interconnected and so individual wire feed control was difficult. This was followed by separating the spools and electromagnetic clutch control of the individual feed rolls. This allowed better individual deposition control for each boat. However these tended to be slow to respond particularly for deposition at high rates and there was a tendency for the wire feed to oscillate between too fast and too slow. Newer drives for wire feeds use stepper motors where the stepper motors have 400 pulses per revolution of the shaft giving an angular motion of less than one degree per pulse. This makes it possible to achieve a speed control to the wire feed of +/− 0.02%. This precision control and faster response time (3) ensures that the pool size can be kept constant and thus minimising spitting.

Even the supply of wire on spools needs to be evaluated. Some suppliers carefully wind the wire layer by layer and it will smoothly unwind. Others do not take as much care and these spools can bind during unwinding which will affect the smoothness of the wire feed which will in turn make the wire feed and pool size erratic and so adversely affecting the product quality.

The aluminium in its molten form is very corrosive and this limits the materials used to make the resistance heated boats. The aim is to have robust boats with the ability to withstand the rapid heating and cooling cycles and with a usable resistance range over the boat lifetime.

The basic boat composition uses a two phase intermetallic mixture of boron nitride and titanium diboride. The boron nitride has high electrical resistivity and possesses good thermal shock resistance combined with ease of machining and the titanium diboride has low electrical resistivity and is easily wet by the aluminium. Some suppliers add a third component of aluminium nitride that is added for its thermal conductivity and electrical resistance performance. The aim of these mixtures is to deliver the best combination of resistivity and corrosion resistance (2, 4–8)

The boats are manufactured using powder metallurgical methods. One route is to mix the powders together with a binder and make a large billet that goes through a heating cycle. The billet is then sliced up to make the desired boats. As the billet is large the temperature seen by the centre of the billet is different to the outside and so the resistivity of each boat will depend on the position in the billet it was cut from. At least one supplier takes a different route and mixes the powder but then compacts and heat treats each boat individually so that every boat is identical. Boats can be purchased with different resistivity values. It is usual to install a set of boats, all with the same resistance value. The aim is that these will age at the same rate and the deposition from each boat will be matched throughout the lifetime of the boats. The lifetime is dependant upon the quality of the manufacturing of the boats. There are some very cheap boats available but the powders used tend to be larger and the coarser structure is less resistant to corrosion and erosion and so the lifetime is shorter. The higher quality boats can be expected to have a lifetime around 15 +/− 5 hours, although this is very dependant upon how the boats are used (9). If the boats are driven hard at high temperature they will have a shorter lifetime than if they are run cooler with a slower wire feed. Over the lifetime of the boats the voltage will drop by 60%–70% and the current requirement increase by 40%–50%. This is as a result of changes to the cross sectional area of the boat due to erosion and chemical changes brought about by corrosion and chemical reactions. Over recent times there has been work done to try to refine the manufacturing to increase the boat lifetimes further as well as save energy by reducing the power required by the boats. The performance of the boats can be plotted as a function of aluminium wire feed rate or temperature and related to the boat dimensions, resistivity, initial voltage, current or power requirements (10,11). This performance change with time means that failure of a single boat presents the problem of how to replace it. It may be possible to replace the boat with one that might have been saved from another set of a similar age and resistivity. Alternatively the failure of the first boat may trigger the replacement of all the boats to start with a completely new set. This may be more expensive but deliver more consistent product. As boats can fail for a variety of reasons there will sometimes be the failure of a boat very early and this makes the automatic replacement of all boats a very expensive option.

The way the boats are operated can affect the lifetime. All manufacturers have recommendations about the heating of the boats for first use and on cooling before venting the system (12). The first time use is important as the powder based boat can be porous and contain volatiles or absorbed moisture. These volatiles need time to migrate to the surface and be burnt off and so stabilise the boats for future use. This first time use temperature rise is slower that that used for subsequent use. Wetting out the aluminium on the boat surface can be problematic and it is typical for operators to cut a length of aluminium wire and place it directly onto the boat to help initiate the wetting before the wire feed is started.

At the other end of the deposition process the boats also need time to cool down before the air is allowed into the vacuum vessel. If this is not done the thermal shock can initiate cracks, or where cracks are already present can open them out and cause premature boat failure. Typically the recommendation is to allow the boats to cool from the bright red/orange/white (viewing colour is always subjective) operating temperature of 1400–1600°C to below the 'black heat' temperature of around 600–700°C before starting the vent process.

Figures 7.6, 7.7and 7.8 show a schematic and then photographs (13) of real boats where it can be clearly see the erosion either side of the cool spot where the wire feed meets the boat. Also clearly visible is the accumulation of large amounts of crud at the ends of the molten pool. As the precipitation and erosion change the cross sectional area of the boat there will be changes to the resistivity and hence temperature changes. There are also chemical compositional changes which change the proportions of conducting and insulator that also change the boat resistivity. This changing performance continues throughout the lifetime of the boat. As can be seen this is a complex process with different chemical and physical reactions taking place. This aggressive corrosive nature of the molten aluminium makes it difficult to find a more stable material that will

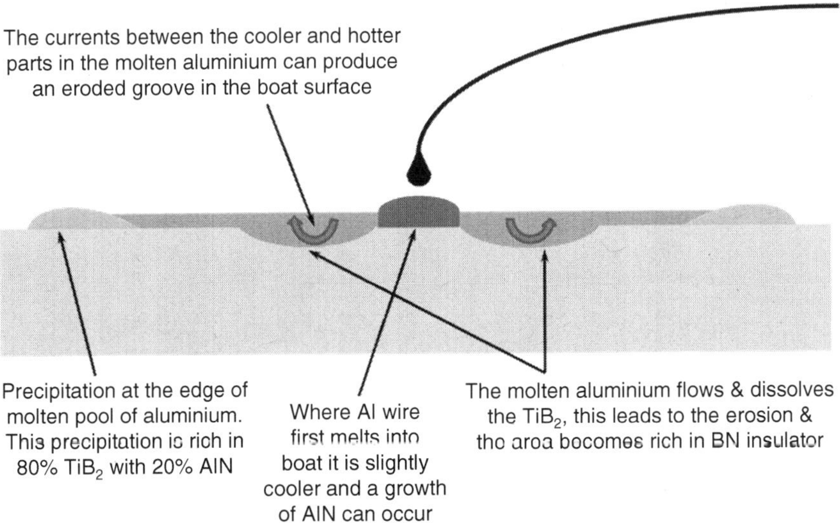

Figure 7.6 A schematic of the chemical and physical changes of a boat during use.

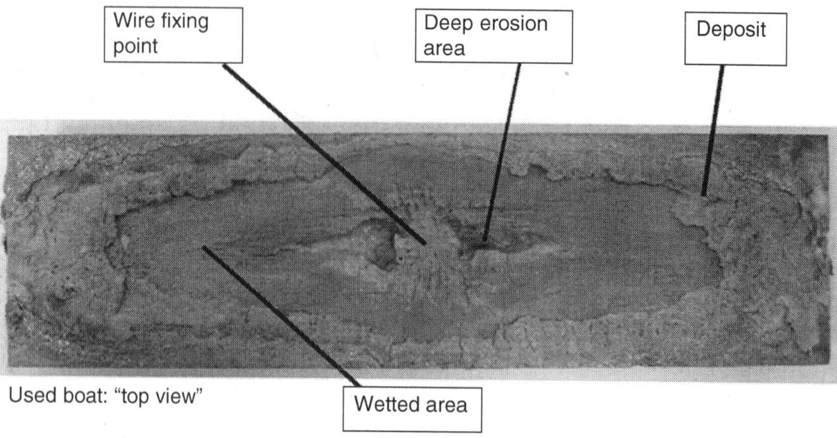

Figure 7.7 A photograph courtesy of Kennametal Sintec Keramik GmbH showing the different areas produced on a used intermetallic boat. Shown at AIMCAL 2007 Conference.

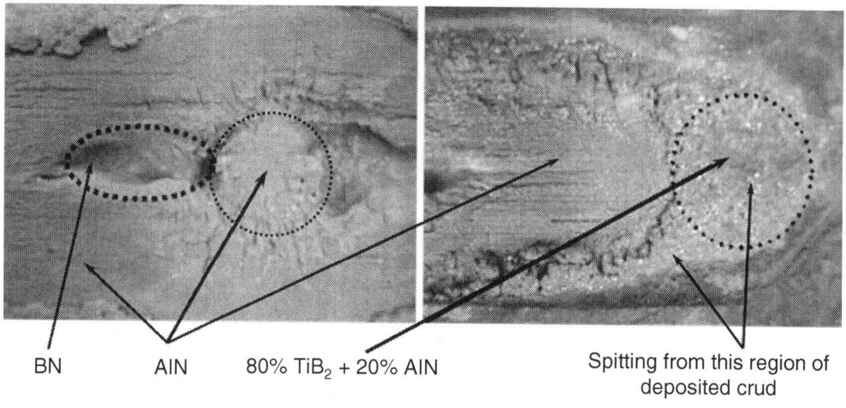

Figure 7.8 A photograph courtesy of Kennametal Sintec Keramik GmbH showing the different areas produced on a used intermetallic boat. Shown at AIMCAL 2007 Conference.

have a longer and more consistent lifetime. The same material has been shown to last longer where there is no wire feed but these sources, which hold the whole material inventory for each deposition run, have a different set of operating problems.

It has been found recently (14) that some of the erosion is due to the binder being attacked and washed away by the molten aluminium. This loss of binder makes the remaining powder fragile and this too can be washed away leading to the erosion. Hence attempts are being made to find a more stable binder which would reduce the erosion and lengthen the boat lifetime.

Rather than allowing the crud to accumulate it is common for the operators to scrape the top surface clean to minimise this crud and so reduce the propensity to spit.

Some boat failures are preventable with the right care and attention. As with other parts of the process the greater the consistency in handling and storing boats the easier and more reproducible the process and product. Although the boats are ceramics they can still be damaged by rough handling. In particular the regions of contact are important. Boats have to expand and so the contacts have to allow for this movement. There are two methods used one is to have the boats a press fit between two copper posts with this location type known as a side clamp. The other that is probably more widely used now is the end clamp where one end is fixed in position and the other is spring loaded. On both of these methods it is important to have the largest surface contact possible. Not all boats are exactly the same size and in order to get the best contact a carbon loaded paper is used that can deform allowing slight variations in the surfaces without losing contact. In particular with the side clamp it allows easy relative movement between the boat and clamp, this is shown schematically in Fig. 7.9.

Figure 7.10 shows some of the common problems found in locating the boats. If there is poor contact between the clamp and boat it can produce a hot spot and this can cause corrosion of the clamp

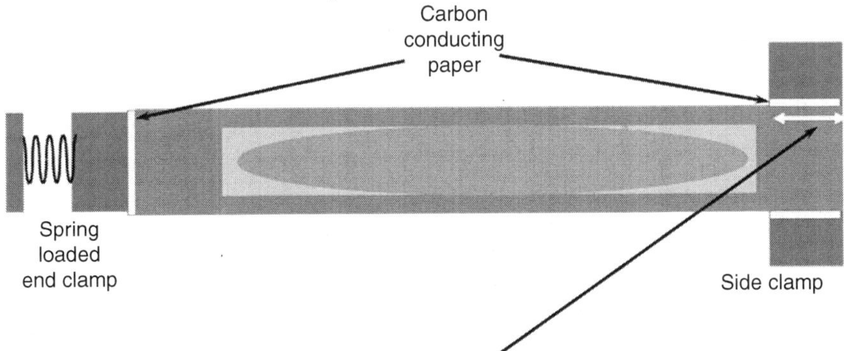

Figure 7.9 A schematic of the use of electrically conducting paper to aid producing a good conduction path when locating resistance heated boats.

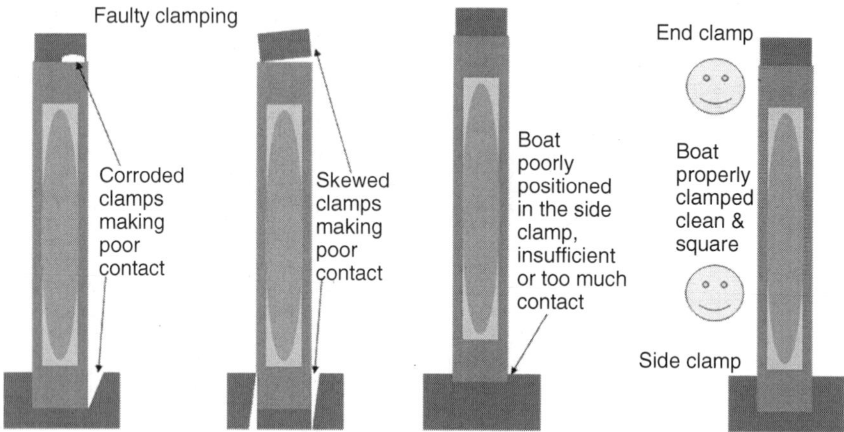

Figure 7.10 A schematic of examples of bad clamping and good clamping.

Figure 7.11 Real examples of boat failures courtesy of Denki Kagaku Kogyo and presented at the AIMCAL Fall Conference 2007 (15).

or boat. If the clamps are not cleaned and made flat this connection problem will continue with each new boat located against this clamp. Similarly rushing to locate the boat and not making sure it is square and in full contact may also produce a hot spot. This makes the control of the boat and deposition rate harder to achieve and this in turn can lead to increased spitting as the molten pool size will be poorly controlled too. Figure 7.11 shows some real examples of overheating, corrosion and cracking of the boat ends that caused an early termination of the life of the boat (15). If operators wear clean gloves and are careful in handling and locating the boats there is less chance of chips, cracks or poor contact.

Not only do the boats need to be located well but they also need to be horizontal otherwise the molten pool will move to the lower end of the boat and skew the deposition from the boat. Not only do the boats need to be positioned correctly but also the wire feed needs to be positioned well so that the wire will consistently feed to the desired spot on the boat as shown schematically Fig. 7.12. For some boat manufacturers they prefer the wire hit the boat at a point one third of the length of the boat but for others they prefer it to contact the boat at the halfway point. This is shown in the photograph in Fig. 7.13. The wire ideally will feed down the centreline and not to one side of the boat. If the wire feed is off centre the molten pool is also likely to be skewed and this too will change the shape of the vapour cloud and alter the deposition profile.

Another key part of the deposition process is to get an even coating over the whole substrate surface. Using roll-to-roll processing has the major advantage that the substrate is in motion in one direction and hence if the deposition sources are run at a constant rate there will be inherent coating uniformity in the direction of the substrate motion. So down the length of the polymer film substrate the coating thickness can be reasonably constant. Across the polymer film it is much more difficult to get coating uniformity. As there are

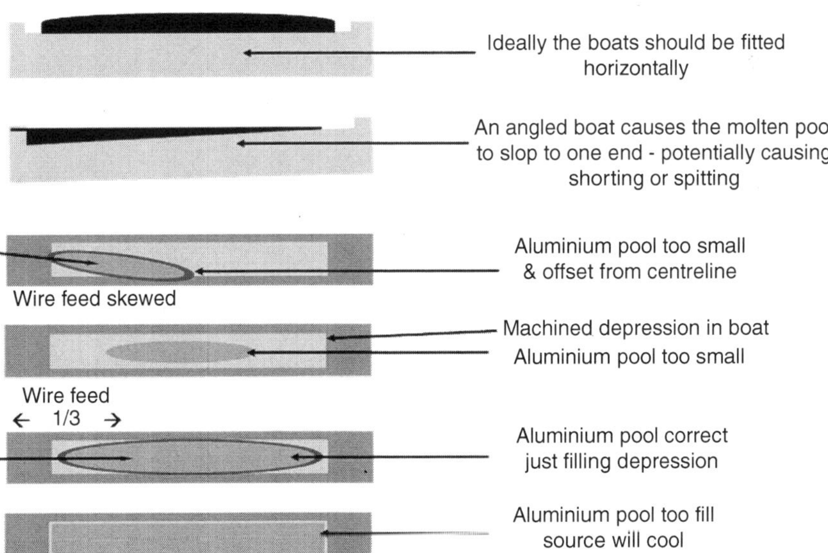

Figure 7.12 A schematic of some common faults seen in resistance heated boats.

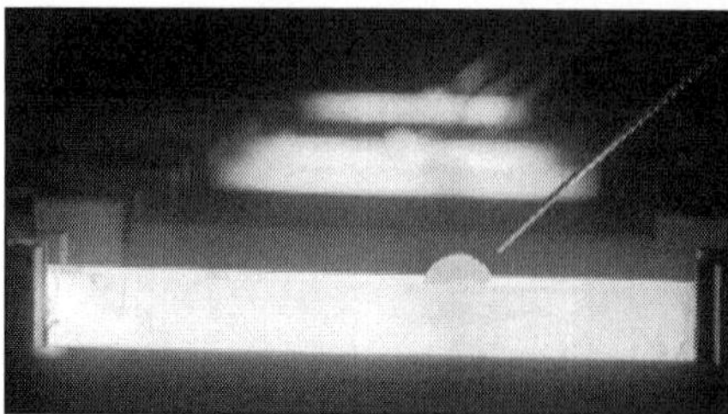

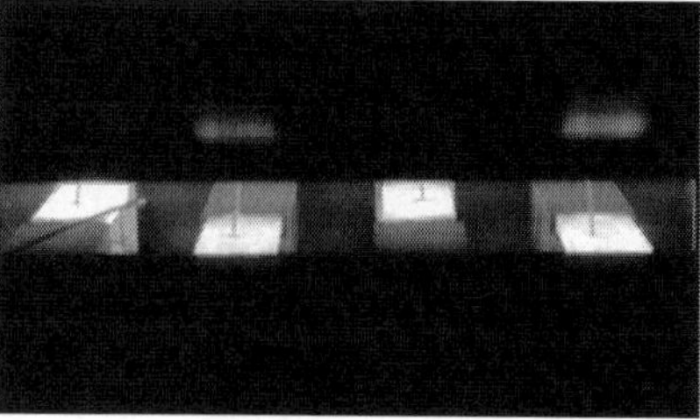

Figure 7.13 Photographs courtesy of General Vacuum Equipment Ltd. On the left it shows a wire approaching the surface of a boat where there is a bead of molten aluminium from a length of wire used as a primer that has yet to wet the boat surface. On the right it shows a series of staggered boats with the wire touching the molten pool almost but not quite on the centreline of each boat.

Figure 7.14 Photographs courtesy of General Vacuum Equipment Ltd. showing the inside of an aluminium metallizer. With detail photographs of wire feed system and arrangement of staggered double set of boats.

multiple boats the aim has to be to get every boat evaporating metal at exactly the same rate which is difficult. The multiple sources and feed systems are shown in the photographs in Fig. 7.14. As every boat could have a slightly different resistivity and the end contacts may have a slightly different contact resistance it means that simply setting the same current to each boat will not guarantee the same boat temperature. Add to this the possible variations in wire feed rate because of spool and feed tube frictions and the quantity of aluminium in the molten pool is likely to be different in every boat. Thus it is important to measure the coating deposited on the polymer film and use this feedback to control the boat temperature and wire feed to keep the thickness constant and as near identical to the deposition from the other boats. Changing the boat temperature will change the size and shape of the vapour cloud (16–19) as shown in Fig. 7.15 and this will also affect the transverse direction coating uniformity which is governed by the overlapping deposition profiles from each evaporation boat as shown schematically in Fig. 7.16.

As the metal coating is conducting it is possible to measure the coating thickness using a non-contact eddy current measurement technique (20–23). These measurement heads can be aligned with each boat, or at the midpoint between each boat, and after calibration the current to the individual boats and wire feeds adjusted to maximise the coating uniformity as shown schematically in Fig. 7.17.

In maximising the uniformity there are various factors that need to be controlled. As mentioned above changing the deposition rate will change the vapour cloud shape and so the amount of overlap between adjacent deposition clouds will change. Some of the remaining variation can be corrected by using profiled shields that reduce the deposition at the high spots. Once the shaped profile

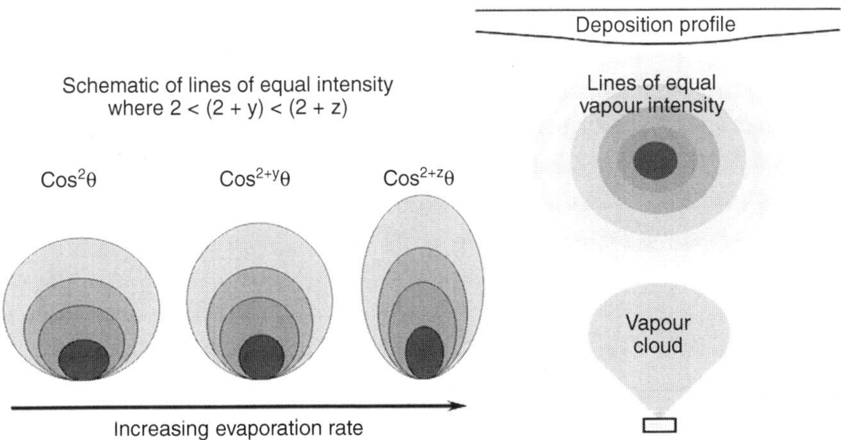

Figure 7.15 A schematic of how the vapour cloud shape changes with increasing boat temperature.

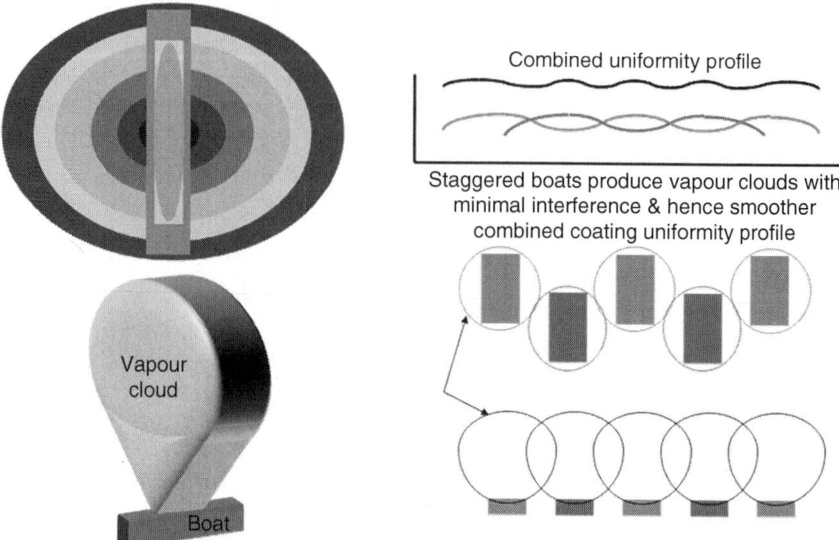

Figure 7.16 A schematic of the vapour cloud profile and how multiple boats can be arranged to give an improved coating thickness uniformity profile.

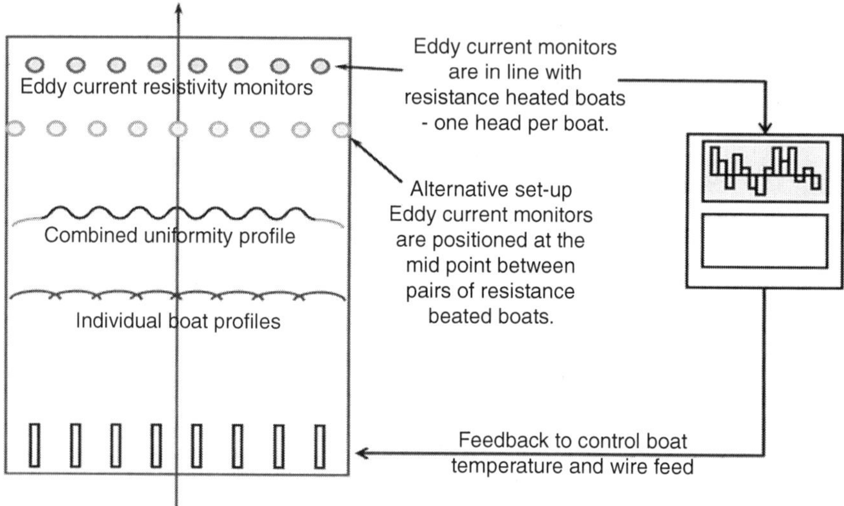

Figure 7.17 A schematic of the options for coating monitoring and feedback control.

shields have been installed this fixes the process somewhat. To run the process faster or slower requires a change in the boat temperature and hence the vapour profile will change which results in the profile shields no longer being optimised.

The difference between high quality good barrier aluminium coatings and poor quality coatings is largely the difference in the fine detail. Choosing high quality boats and wire, storing them in a controlled environment and then handling them with care will pay back in longer life for the boats with fewer failures and reduced spitting. If you add to this the aim of having the most stable process possible with the fewest changes in power, boat temperature and wire feed rate will also minimise the spitting. All of this will aid producing more consistent coatings of higher quality with fewer defects that can reduce the barrier performance.

There has long been a requirement for transparent barrier coatings of an equivalent performance to the aluminium metallized coatings (24–33). There has been no problem about being able to produce such coatings but the problem has been how to deposit them as cheaply as the aluminium metallized coatings. The transparent coatings tend to be oxides which have much higher melting points than aluminium and so required a different deposition source to deposit the coatings and these tended to be higher cost systems. This was usually coupled to slower deposition rates and so the cost of the coatings started at more than 10x higher than aluminium but has progressed down to somewhere around 2x–3x higher in cost depending on the process and materials.

As aluminium oxide is transparent and provides the natural barrier for aluminium metal it has been one of the materials evaluated for barrier coatings. To get a low cost process one metallizing company found that adding oxygen to a standard metallizer could produce an oxidised aluminium coating (34–37). This is not necessarily as easy as it sounds as the oxygen can attack the aluminium molten pool and slow the deposition rate (38) and it can also strike up an unwanted plasma in the vacuum system that can give problems to the power control. However when it is suitably controlled it can provide a low cost transparent barrier material. Others have looked at this process and now there are several proprietary processes that compete for this market an example (39–41) of which is shown in Fig. 7.18. Also by using the plasma and enhancing it the coating density can be increased reducing the defect size which in turn improves the barrier performance (42).

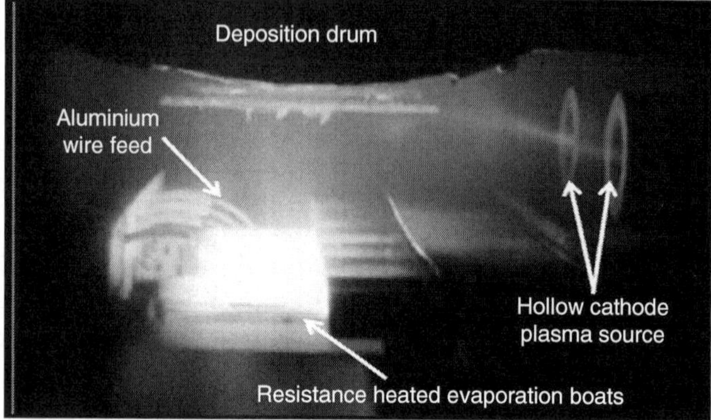

Figure 7.18 A photograph from the inside of a development plasma enhanced deposition system courtesy of Fraunhofer FEP Dresden as presented an AIMCAL Conference (39).

7.2 Plasma Enhanced Chemical Vapour Deposition (PECVD)

Other transparent barrier materials directed at food packaging applications include silica, carbon, melamine with other compounds being used for the ultra barrier applications. Mostly the roll to roll vacuum system manufacturers compete with one of two other deposition technologies for this large market opportunity. The two competing technologies are plasma enhanced chemical vapour deposition (PECVD) and electron beam deposition. The PECVD process requires a suitable precursor liquid or gas that can be decomposed in a plasma and condensed and deposited onto the moving web in a suitable chemical form that produces a transparent barrier coating (43–47). The PECVD process is simple in concept but producing a high rate deposition process that reproducibly produces the same barrier product has proved difficult. The flow of the precursor gas can be controlled well using mass flow controllers and the additional gases to maintain the plasma and deliver the correct coating stoichiometry can similarly be fed in through mass flow controllers. The difficult part is to produce and maintain the plasma characteristics over the length of the deposition run. The most popular precursor chemical is hexamethyldisiloxane (HMDSO) which with the addition of helium and oxygen will produce a silica coating. The by-product

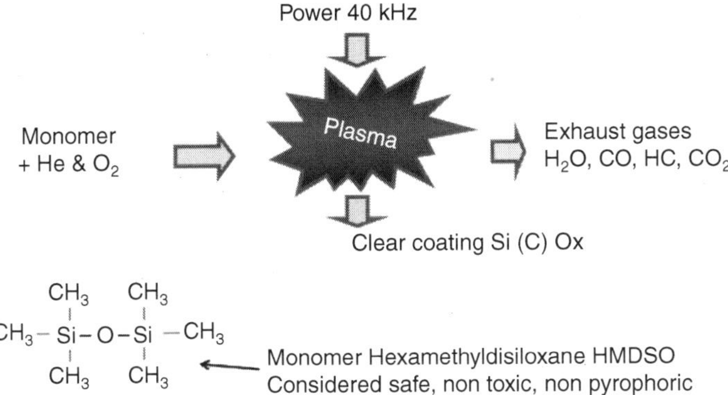

Figure 7.19 The basic chemistry of PECVD deposition of HMDSO.

gases are gases such as carbon monoxide, carbon dioxide and water which are all easily pumped away as per Fig. 7.19.

The plasma density can be enhanced by using magnetic fields to reduce the electron loss which speeds up the decomposition process. Where some of the difficulty arises is that the silica coating is non-conducting and the coating not only deposits on the substrate but will also deposit onto all other surfaces which includes any earth or anode. This dielectric coating then reduces the effectiveness of the anode or earth and the plasma performance declines as too does the deposition rate and control of coating stoichiometry.

This disappearing anode problem has been counteracted by using hot anodes that run at a high enough temperature to prevent any condensation of the vapour that would be converted into a coating. As the gas can be fed into the system in any direction there is no restriction on the deposition orientation and so the geometry of the system can allow a larger proportion of the deposition drum to be used for deposition. The limitation of this is that as the incoming gas is decomposed and condensed onto the polymer film there is less active gas available to deposit onto the next part of the film. Also the by-products need to be removed to limit their incorporation into the coating or from contaminating the process. In Fig. 7.20 it shows the gas being fed into a centre point of the plasma which is concentric to the film around the deposition drum. In this way the gas only passes part way around the drum before the exhaust gas is removed. The spacing of the plasma volume can be important as if the distances are too large and under certain conditions, it is possible to produce powder that can be a source of contamination

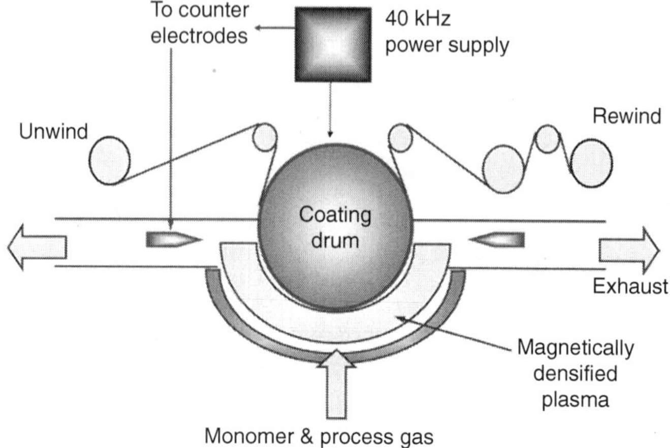

Figure 7.20 A schematic of the PECVD process for producing silica coatings.

and pinholes in the growing coating. The latest system to have been built was equipped with three deposition drums but even so this still could only deposit a barrier coating at approximately half the deposition speed of a modern aluminium metallizer. This three drum system had to be bigger to incorporate three deposition drums and hence was a significantly higher cost than the equivalent metallizer. Thus the cost of the transparent coating is still of the order 2x–3x the cost of aluminised polymer films.

The coatings produced by PECVD may differ from what might be expected of the same coating produced by a physical vapour deposition. The deposition of silica by PECVD can produce coatings almost identical to those produced by electron beam deposition but by changing conditions differences can be introduced. It has been shown that the amorphous network structure of silicon and oxygen atoms can be modified to also include carbon atoms (48,49). It is possible to include up to 20% of carbon atoms which are thought to sit in the holes in the amorphous network which effectively blocks up some diffusion routes through the silica structure and so improves the barrier performance. The addition of carbon does not add colour and the silica coatings remain transparent and colourless.

7.3 Electron Beam Evaporation Sources

The major competing technology for the PECVD uses a large electron beam deposition source. The electron beam deposition source uses a source of electrons that uses magnets to focus the stream of electrons onto the target source material that is heated by the electrons as shown in Fig. 7.21. If sufficient electrons hit the source material the material can be heated to a temperature where evaporation or sublimation occurs.

In the same way that multiple resistance heated boats can be used to coat wide polymer webs with overlapping deposition vapour clouds so too can multiple small electron beam guns be used in the same configuration (50) as shown in Fig. 7.22. The sources do not tend to suffer from spitting as the resistance heated boats do but

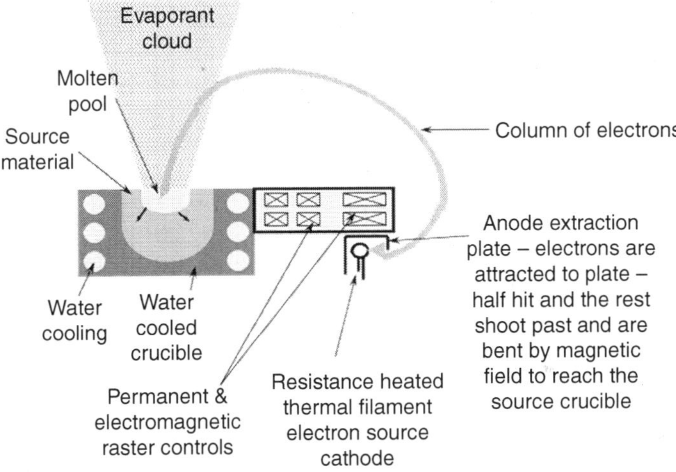

Figure 7.21 A schematic of a small electron beam gun.

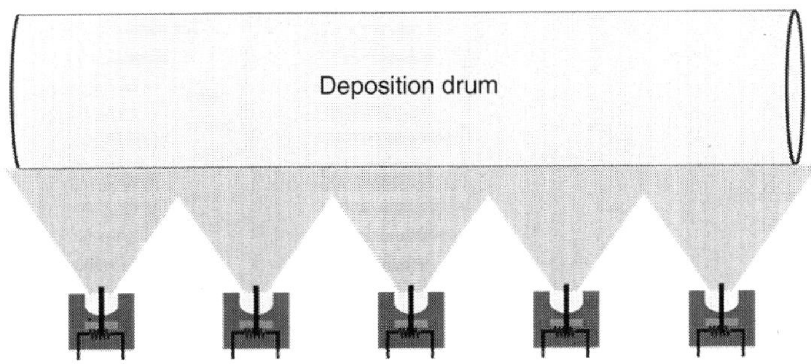

Figure 7.22 A schematic of the use of multiple small electron beam guns for roll-to-roll polymer film vacuum deposition.

can still damage the substrates by other methods. The first potential problem is during the initial heating process where the fresh charge of material has to be heated for the first time. Any defects in the material such as voids or moisture can heat up faster than the bulk material and the thermal expansion of the air or water can cause a mini explosion of the material. This can be violent enough to eject a sufficient amount of material to have to stop the process. The material ejected can sometimes be sufficiently hot that if it hits the substrate it will damage the substrate either by sticking to the substrate or burning holes in the polymer film. If the material heats up evenly and melts but the electrons are too well focussed the beam can drill a hole into the molten material and as the liquid falls into the hole it can eject the liquid rapidly and again this can damage the substrate. However once heated and stabilised this type of source can be run with few problems (51).

Using the same basic process but by using a much more powerful electron source it is possible to use a single electron beam gun to heat a linear source of up to around 1.25 m wide (52,53) as shown schematically in Fig. 7.23. In this way a couple of electron beam guns can deposit a coating width of 2.5 m which, for aluminium deposition, would take of the order of 25 resistance heated boats.

One benefit of electron beam heated deposition sources is that switching between depositing materials is easy as it only requires

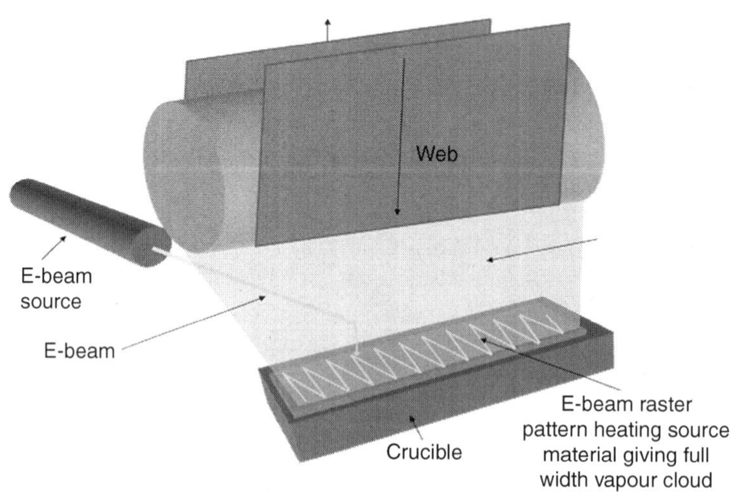

Figure 7.23 A schematic of a large electron beam deposition process.

the crucible to be cleaned or exchanged and a different material loaded and a different coating can be deposited. This means that it is easy to use a system like this for the deposition of aluminium opaque barrier materials as well as silica, alumina or other transparent barrier materials. This type of source is more expensive than the resistance heated source type and has a high deposition rate and so the cost per square metre of material reflects this difference in capital cost.

The deposition from this type of source is different from simple resistance heated boat evaporation in that the vapour evaporating from the source surface has to pass through the incoming electron beam. A proportion of the vapour will be ionised. There will also be the production of secondary electrons from the source (54). The secondary electrons are energetic enough for some to be implanted into the polymer film and this can cause electrets to form in the polymer that can cause the film to block once rewound. This problem is prevented by the use of a secondary electron trap (55–58). The substrate then sees a heat load from the latent heat of condensation, radiant heat from the hot source and heat from the energetic proportion of the depositing vapour. The bombardment of the depositing coating also improves the adhesion and density of the coating. This bombardment can be increased by the use of an additional plasma that is positioned between the source and substrate (39–42,59,60). This can significantly densify the coatings but there is also a danger that the higher density coatings also have a higher intrinsic stress that can make downstream handling problematic. This type of process, shown in Fig. 7.24, also allows the evaporation of metals and using the additional plasma and the input of reactive gas the conversion of the metal to some other compound such as titanium to titania or titanium nitride. Again this type of source provides a wider range of materials that can be deposited which can have commercial advantages. The high energy of the electron beam can change the stoichiometry of the coating from that of the source. This can require some gas to be added to help bring the stoichiometry back to the required proportions. Variations in stoichiometry can give rise to variations in barrier performance (61–63) and also optical performance. Again the stability of the source operation and any gas feed is critical to delivering high quality barrier coatings.

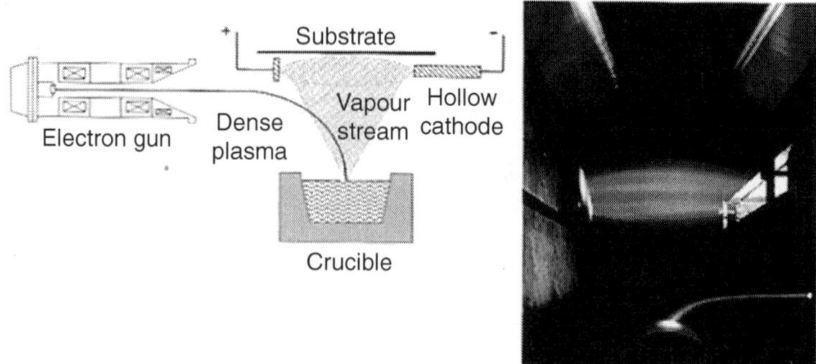

Reactive EB-evaporation and hollow cathode plasma-activation

- Deposition of Al_2O_3 by evaporation of alumina
- Deposition of Al_2O_3 by evaporation of Al
- Deposition of SiO_x by evaporation of silica
- Deposition of TiO_2 by evaporation of Ti

FEP - Dresden

Figure 7.24 An example of using an additional plasma courtesy of Fraunhofer FEP Dresden (64).

7.4 Induction Heated Evaporation Source

A source that found favour in Japan for the deposition of silica for transparent barrier materials was the induction heated evaporation source as shown in Fig. 7.25. This source was generally used, as seen in the photograph, as a series of circular sources and so used the same principle of overlapping deposition vapour clouds to provide suitable coating uniformity.

The way the source works is that a high current medium frequency RF source (250 Hz–25 kHz) is passed through a water cooled circular coil inside which the crucible is located. The eddy current produced by the high current passes down the centreline of the circular crucible and the changing polarity creates vibration in the molecules of the source material which heats up. This type of source can be used for a wide range of materials (65–69). This method of heating is more effective on magnetic or higher resistivity materials than it is on non-magnetic or lower resistivity materials.

The induction heating sources have similar problems to electron beam heating sources in that any trapped gas or porosity can result in the gas or water expanding too quickly and produce spitting or explosions in the source during heating.

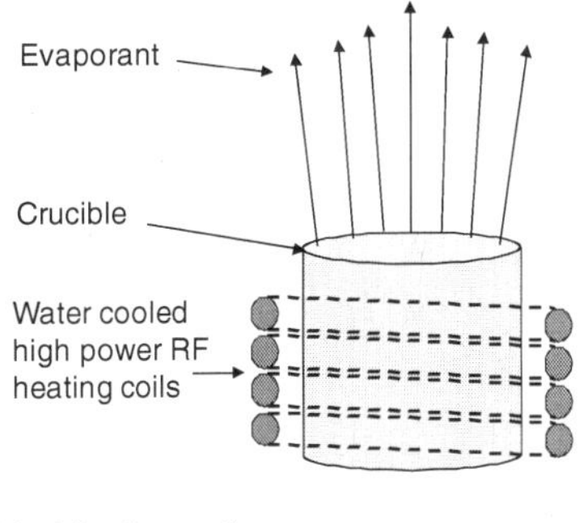

A high current passes through the heating coils. This causes an induced magnetic field inside the coils where the RF supply (~1 kHz) produces a rapidly reversing polarity magnetic field causing the source molecules vibrate & heat up.

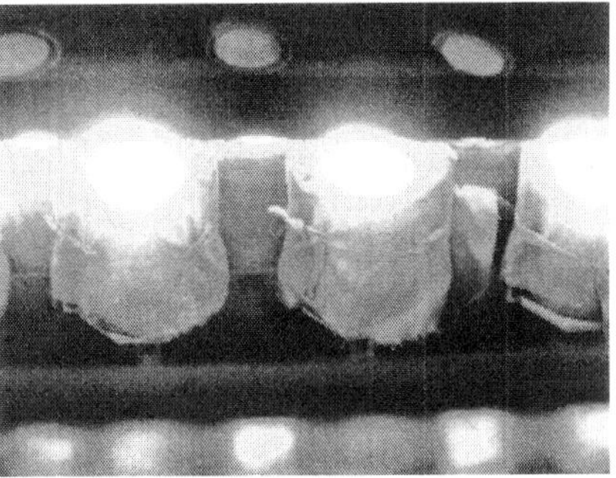

The heating coils are hidden by the insulating fabric that is used to both reduce the radiant heat load from the source but also to facilitate cleaning of the source

Figure 7.25 A schematic and photograph of an induction heated evaporation source.

There has not been a clear advantage of this type of source over the electron beam source. The electron beam source is a higher cost source but can provide superior coating uniformity.

7.5 Magnetron Sputter Deposition Sources

Sputter deposition stands out as the source material stays as a solid as opposed to the majority of evaporated materials that become molten. As the source material, known as the target, remains a solid it allows the deposition sources to be used in any orientation around the deposition drum that also allows multiple sources to be used around the same deposition drum. The way that sputtering sources work is that the target is arranged as the cathode and is used as one of the electrodes to strike a plasma (70–73). The ions in the plasma bombard the negatively charged cathode surface and if the bombardment is sufficiently energetic will eject some of the target material, known as sputtering. This ejected material will deposit onto the adjacent polymer film on the deposition drum. This process, by comparison with evaporation, is very slow often of the order 1000x slower in deposition rate. The basic process, known as direct current (DC) diode sputtering, operates at around 2 kV. The plasma density is low and the electron loss to the walls of the chamber is high and so the current carrying capacity of the plasma is limited. To improve the deposition rate the plasma density had to be improved and this was done by adding additional electrons by using a specific electron source (74,75) or preferably, by using magnetic confinement (75–78). The electrons are constrained by the magnetic field lines and hence are not lost as rapidly to the walls of the chamber. Also as the electrons spiral under the influence of the magnetic field they travel much further and this also increases their chances of undergoing ionising collisions. There is both an electric field and a magnetic field that will affect the motion of the electrons. The electrons will spiral around the magnetic field lines and as they approach the negatively charged target material will be reflected and spiral back around the field lines until they undergo a collision. When they undergo a collision they can either be knocked closer or further away from the target surface. In addition to spiralling around the field lines the electrons will also precess around the target surface under the combined influence of the crossed magnetic and electric fields. This is shown schematically in Fig. 7.26. Where the magnetic field is parallel to the surface of the target material there will be the maximum erosion of the target by sputtering. The magnets are arranged so that the precession of the electrons form a closed loop over the target surface. This can be from a planar surface such as circular or rectangular target but so long as there is a continuous racetrack any target shape can be sputtered of which that most widely used alternative to planar targets are cylindrical targets. The target size is usually made to be slightly wider than the polymer film width to give a high coating uniformity. This erosion profile of the target is known as the racetrack and the shape and depth is affected by the magnetic

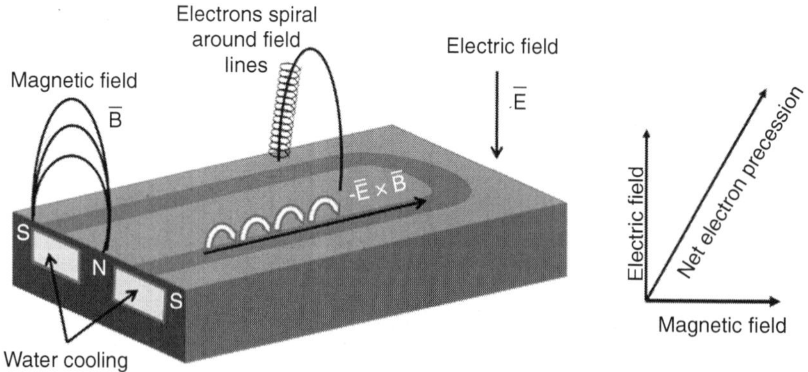

Figure 7.26 A schematic of a planar magnetron sputtering source.

design of the magnetron. This increased plasma density allows the plasma to be run at around 500V but with much higher currents and as the sputtering rate is proportional to power the magnetron sputtering sources can produce deposition rates an order of magnitude faster than diode sputter sources. The magnets can be positioned in a variety of different ways to optimise different features such as deposition rate, material use efficiency or minimise heat load (79–90).

As there is no high temperature source the heat load seen by the polymer substrate has different proportions for each of the different contributions. The target is bombarded by the ions from the plasma and not every impacting ion will eject an atom from the surface with many impacting ion simply transferring some of their energy as heat into the target material. To keep the target surface a solid the back surface of the target material is cooled. However the front surface is above ambient temperature and so will radiate some heat to the substrate although this will be considerably smaller than for an evaporation source. The latent heat of condensation will be the same for the same thickness of coating. The more significant heat load is from the energetic material that is deposited. This energetic material will improve the adhesion of the coating as well as increasing the density of the coating deposited. This concurrent bombardment of the depositing material can produce a coating with fewer defects, for the same material and coating thickness, than for a coating deposited by simple evaporation. The design of the magnetron sputtering source can affect the heat load seen by the substrate. The design extremes are known as balanced and unbalanced magnetrons (91–97) and this is shown in Fig. 7.27. Using an optimised

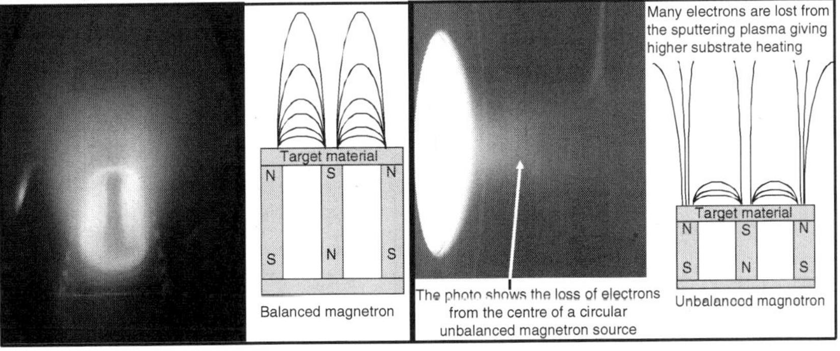

Figure 7.27 Details of balanced and un-balance magnetrons.

magnetic design where the amount of magnetic material is in balance the electron loss is minimised and this includes minimising the loss of electrons to heat the substrate. By comparison if the magnetic materials are mismatched it is possible to encourage the electrons to hit the substrate or growing coating and this energy can be used to modify the growing coating structure.

Most of the barrier coatings that are vacuum deposited use polymer substrates that are temperature sensitive and so for the bulk of these barrier coatings the magnetron sputtering source design needs to be of the balanced type. In Fig. 7.27 the right hand photograph is of an un-balanced source which had sufficient energy in the column of electrons emanating from the centre of the magnetron to carbonise a polyester film in a fraction of a second. Even when using higher temperature substrates this source of substrate bombardment needs to be used with care if it is not to do more damage than any improvement it can give to the coating.

It is possible to produce high quality coatings by magnetron sputtering but there can be problems with some deposition processes. As sputtering takes place from a solid surface it does not matter what the melting temperature is of the target material and so ceramic oxides can be sputtered as easily as metals although the sputtering rates may be slower for the oxides than the metals. As the process is already slow by comparison with evaporation it is preferable to maximise the possible sputtering rate. To produce silica or alumina for transparent barrier coatings it is possible to sputter the metal and using a reactive deposition process to convert the coating to the oxide. There are a number of possible problems that are more noticeable in reactive magnetron sputtering than when sputtering simple metal coatings. The sputtering process is not perfect. Some of the sputtered material may return to the target surface. Where the sputtering of the surface is faster than the re-deposition there will be erosion of the target but on the edges of the racetrack the re-deposited material will build up. This build up of non-conducting material can be a source of arcing as the non-conducting material will charge up and may short to the plasma. Any arcing will be hard to extinguish as all the current will have been focussed into the arc making the current density huge and the target surface may well have melted. A schematic of arc formation is shown in Fig. 7.28. This can lead to some incandescent material being ejected. The secondary electron emission will be increased at the arc and so even if the power supply is switched off and then switched back on

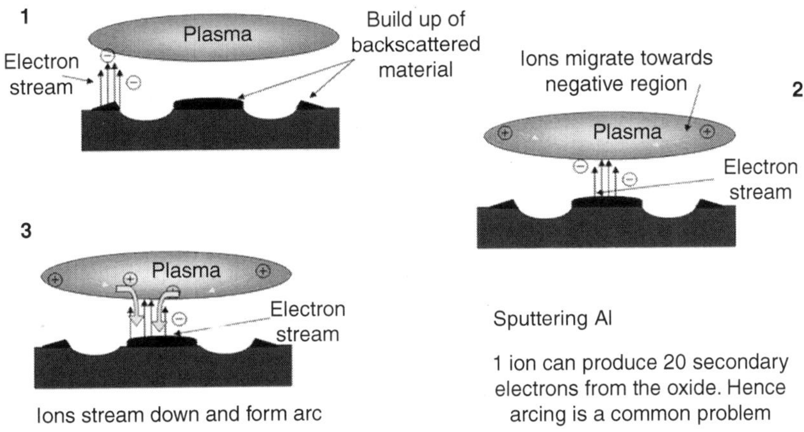

Figure 7.28 Schematics of the formation of an arc on a magnetron sputtering target.

the arc is likely to reform as the electron emission from the hot spot will be the easiest strike point for the plasma. Even if the spot is cool the roughness introduced by the arc can also cause an increase in electron density and arc initiation. This has led to the development of arc suppression circuitry for the power supplies that can sense when an arc is forming and once sensed the power supply reverses polarity to speed up the quenching of the arc. This minimises the energy and so heating of the target and enables the plasma to be re-established more quickly (98–120). In developing the power supplies one of the options is to switch the polarity of the power supply proactively. This has the effect of discharging any surface charging and if the frequency is high enough this will prevent most of this type of arcing. This is shown schematically in Fig. 7.29. During an arc the sputtering stops from the bulk of the target and there is a much more rapid deposition from the arc area or, if the arc is quenched well, there is an interruption in the sputtering. This interruption can cause a variation in the coating thickness. Where the deposition is reactive this changing deposition can be amplified as the proportions of metal and oxygen are disrupted.

Another source of potential problems is that of the disappearing anode. Any stray depositing material will cover shields and the vessel and over a long period of time this can produce an insulating surface. Some vacuum systems use the vacuum vessel, which is allowed to

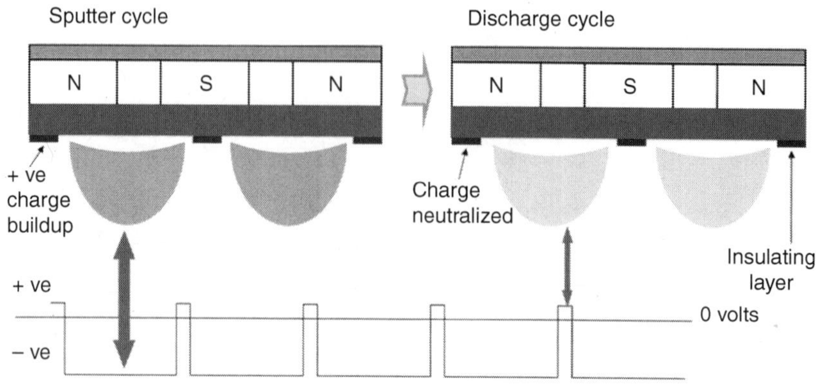

Figure 7.29 A schematic of the use of a pulsed power supply to reduce the arcing from a build up of insulating material on the target surface.

float with respect to earth, as the anode to the sputtering plasma, others will have a specifically powered anode (121–124). Either of these anode types, once coated with an insulating layer, will no longer continue to work as the plasma anode and so the plasma may either extinguish or move as the anode moves around to the next most suitable position (125–131). Thus it is essential to know which surface is acting as the anode to the sputtering source and to make sure it is kept clean and conducting. In some systems this is done during deposition by heating the anodes to prevent any deposition onto the surface.

As deposition times become longer this can become an increasing problem. To counteract this it is possible to use dual cathodes with a switched power supply to ensure an active anode at all times. What happens is that for one half cycle one cathode is sputtered as normal and the other acts as the anode and for the second half cycle the polarity is switched and so the roles are reversed with the original cathode now acting as the anode and the original anode acting as the cathode and being sputtered. As for each half cycle the target surface is being cleaned it will always be clean when required as an anode.

This process can be taken one step further to eliminate both of the above two problems. This uses a rotatable magnetron sputtering source (132–138). Instead of a planar target this uses a cylindrical target which is arranged so that it rotates and so the whole

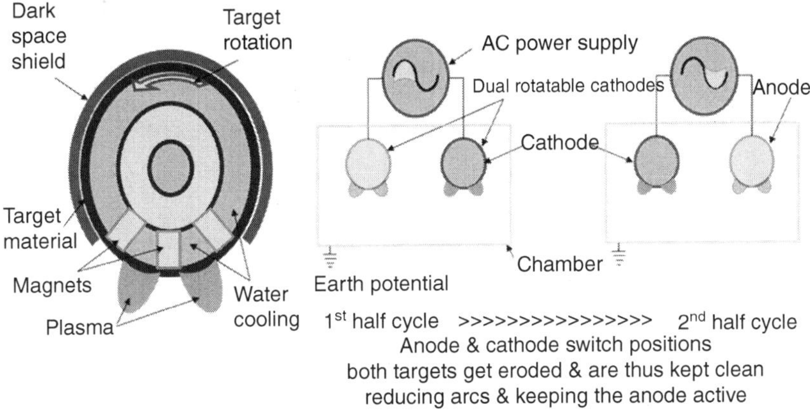

Figure 7.30 A schematic of a rotatable magnetron sputtering source & a dual rotatable magnetron source arrangement.

circumference of the cylindrical target surface passes across the front of the magnets and is sputtered as shown schematically in Fig. 7.30. This movement of the target material to sputter the whole circumference prevents any of the backscattered material building up and becoming a potential source of arcing. If this full surface erosion is coupled to using two cathodes in dual mode, as per Fig. 7.30, then the amount of arcing can be minimised and in so doing maximising the integrity of the depositing coating and the barrier performance.

Where the sputtering process uses a reactive gas the gas cannot be limited in where it goes and it will not only oxidise the depositing layer but also will try to oxidise the target too, as shown schematically in Fig. 7.31. The target surface will also oxidise over time whilst the sputtering source is not being used and so the first action when switching on any source is to first sputter clean the surface to bring the target back to clean metal.

As there may be a different secondary electron emission from the oxide surface than from the metal surface the plasma characteristics will be different for the oxide or metal. The oxide will be cleaned up first at the point of maximum erosion in the racetrack and more slowly towards the edges of the racetrack and not at all from areas on the target where there is normally a build up of backscattered material. If we look at the graph and schematics of the target oxidation state in Fig. 7.32 we can follow the progression of target oxidation from 'A' through to 'E'. At point 'A' the

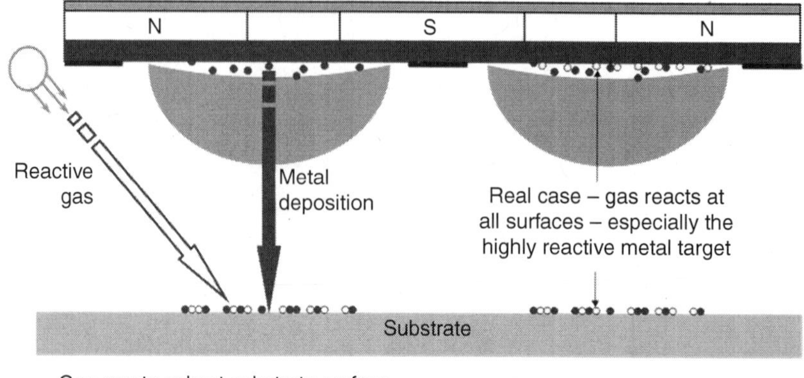

Figure 7.31 A schematic of the reactive deposition process.

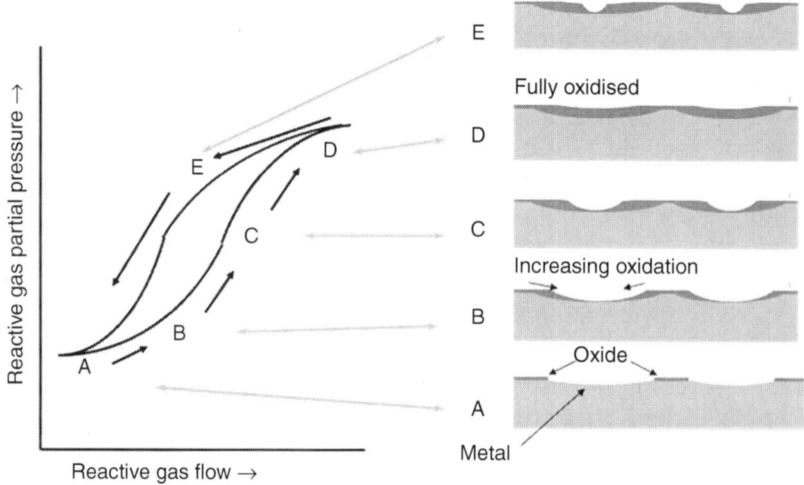

Figure 7.32 A magnetron sputtering hysteresis loop relating to target oxidation.

target surface is predominantly metal and there is no reactive gas present. If we now introduce some reactive gas flow there is some increased oxidation of the target which starts to encroach on the racetrack. This encroachment reduces the metal sputtering rate and so reduces the requirement of the target for the reactive gas and so the partial pressure will rise and we reach point 'B'. If there is a further increase in gas flow the same process will happen again and we will move to point 'C'. Depending on the different sputtering rates of the metal and oxide will depend on how much partial pressure rise will be caused by how much encroachment. This process can become unstable and a small gas flow increase can result in a large excess partial pressure rise. Eventually there will be a direct correlation between the gas flow increase and partial pressure rise as the target is fully oxidised as shown as point 'D'. If the reactive gas flow is decreased the fall in partial pressure does not fall back along the same path on the graph as for the gas flow increase. With the reduction in reactive gas flow the oxide from the target surface will gradually be sputtered away and as soon as the metal is exposed there will be a fall in partial pressure as the increased metal sputtering rate getters more reactive gas. This different rise and fall of partial pressure, or plasma voltage which shows a similar characteristic, is known as the hysteresis loop. The fact that there is a loop is important as it requires that to get the same sputtering rate and reaction process the particular point on the graph needs to be approached from the same direction every time. Hence the target needs to be cleaned up first before feeding in the reactive gas. There have been many studies on the stability of the reactive sputtering process (139–154) and they have shown the balance of pumping between the system pumps and the growing coating acting as a pump as a key factor. There have been various strategies used to try to improve the process control by better controlling the reactive gas flow introduction or by adding some high surface area collection shields as well as closely monitoring the plasma content and using this information for feedback control (155–165). The use of plasma emission spectroscopy to determine the plasma content by identifying the ionised species present in the plasma and their relative proportions (166–173) enables the fine tuning of either the metal sputtering rate or reactive gas flow to optimise the ratio to produce the desired stoichiometry of the depositing coating.

As with all deposition processes the aim is to get consistency and stability into the process and the optical emission spectra is one of the better options for doing this as the information is real time as it takes information from the depositing flux directly. Taking the information is only part of the control process. There are various time constants that are involved and with larger systems these are often longer and so transferring processes is not necessarily a simple matter of scaling. The distance between the gas flow meter and the introduction of the gas into the system will slow down the response time of the process. This can be made worse by the design of the gas manifold (174–177). Particularly in reactive processes the coating uniformity and stoichiometry becomes sensitive to the uniformity of pumping, gas introduction and sputtering uniformity. As polymer films become wider the uniformity and symmetry about the film centreline also becomes more important. The coating stoichiometry relies on the correct ratio of metal to reactive gas atoms or molecules being present on the substrate surface at the right time and with the right energy. If there is not uniform pumping across the whole film width then there will be a pressure variation which will affect the mean free path and deposition rate of the metal and the partial pressure of the reactive gas which will lead to variations in the coating stoichiometry. If there is a pressure gradient then the plasma emission spectroscopy will not be able to control the process well as what works well for one position will not be good for everywhere else. With wide film widths then it is common to split up the gas flows which in turn requires separate emission signals for different positions on the sputtering source. All of this becomes easier if the system was designed to deposit the coatings by a reactive process. Retrofitting a reactive process into a system designed only to deposit metals can be more difficult as the deposition process for metals is so much less sensitive to small process variations.

For some oxide or compound materials it is possible to overcome a number of these problems by sputtering the oxide or compound directly using a suitable oxide or compound target. With non-conducting materials this can be done using radio frequency (RF) sputtering (178–187). The sputtering plasma contains ions and electrons. The electrons are very light and move quickly whereas the ions are much heavier and move more slowly. In an RF field the ions are slow to change direction and so when the RF is applied to the target the electrons will be attracted to the target for half the cycle and the ions for the other half cycle. Initially when the power is first applied the target is at zero potential. When the target is positive the electrons will be attracted to the target surface and the ions repulsed. Then as the target switches to being negatively charged the electrons in the plasma will be repulsed and the ions attracted to the surface. However the ions are harder to get to change direction and so fewer ions will reach the target surface than electrons do during the other half cycle. This difference means that the target surface will progressively become charged negatively and as the target is non-conducting the surface charge will not be dissipated.

Once there is this negative charge present there will be an increased ion bombardment that will start to sputter the surface. The amount of negative charge will reflect the different bombardment rates between the ions and electrons and they will become in balance at some negative potential. As the power is effectively only applied for half the cycle the sputtering rate will be much lower than for a similarly sized DC power supply. RF power supplies are also more expensive and there can be difficulties in matching the RF power into the plasma consistently. Ceramic targets can be more expensive to manufacture than metal targets. Ceramics may also be prone to internal defects and be brittle making them prone to target cracking. Cracks and defects once exposed can be sources of arcing leading to coating non-uniformity and possible pinholes. Thus there are many reasons why not to use RF magnetron sputtering for oxides. However there is no hysteresis loop for RF sputtering ceramic targets and so it is possible to more easily achieve process stability. Thus for many production processes it is easier to accept the higher cost and slower rate but to RF sputter from an oxide target.

7.6 Atomic Layer Deposition (ALD)

The technique of atomic layer deposition (ALD) has been around for many years. The process consists of introducing a precursor gas that will attach to all surfaces as a monolayer. The excess gas is pumped away and a second gas introduced that also condenses and converts the first precursor into the desired coating (188,189). The excess second gas is pumped away and then the whole process is repeated again to deposit a second monolayer. This sequence is repeated as many times as necessary to deposit the desired thickness. One of the attractive features of ALD as a process is that the coating deposited is conformal with the substrate surface. Conformal meaning that the coating follows the surface contours so that there is the same thickness of coating over the whole surface irrespective of surface roughness or defects. This makes this coating process interesting as it looks as if it has the potential to reduce the number of coating defects that are caused by substrate effects.

Recently the ALD process has been developed to make it possible to deposit coatings onto polymer film substrates (190, 191). These two papers describe two quite different approaches to the design of equipment for the roll-to-roll ALD.

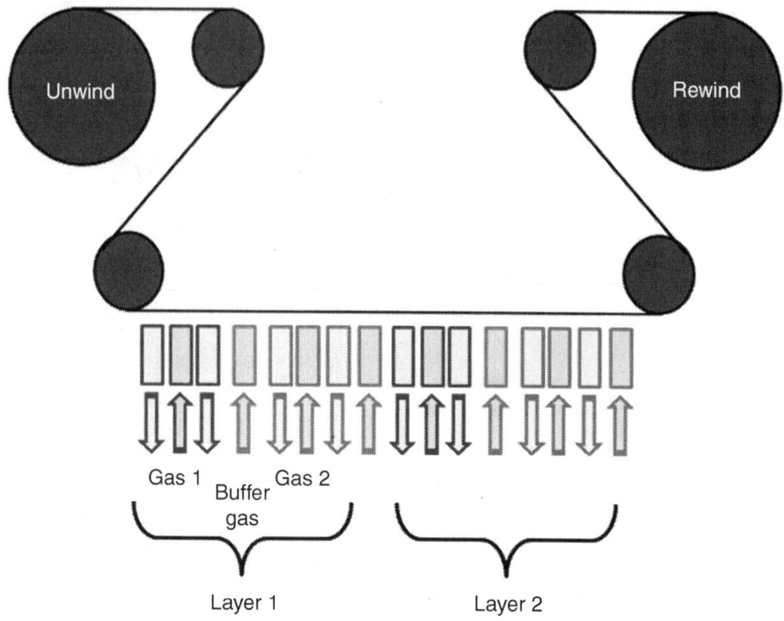

Figure 7.33 A schematic of a multiple source roll-to-roll ALD process.

The first of these two designs has taken the idea of making a compact delivery system that could be multiplied up to make it easier to build up the coating thickness. This is shown schematically in Fig. 7.33 where the polymer film is wound through and sequentially exposed to the first precursor gas, then the excess is pumped away and this is followed by a buffer gas and then the second gas is introduced followed by the excess being pumped away. Using the right manifold design the gas introduction can be made to be uniform across the full width of the film. This design came from the semiconductor end of the industry and the work was designed at producing multilayer electronic coatings. Using multiple deposition heads and gas delivery systems makes this a high precision but higher cost system than the other approach which is shown in Fig. 7.34.

In Fig. 7.34 the polymer film is wound in a serpentine geometry so that the film repeatedly passes through a zone with gas 1 then through a buffer zone and then through gas zone 2 and back through the buffer zone. This cycle is repeated multiple times to build up the coating thickness. The simplicity of this system is that there is only a single feed for each gas and this makes the cost lower. This design

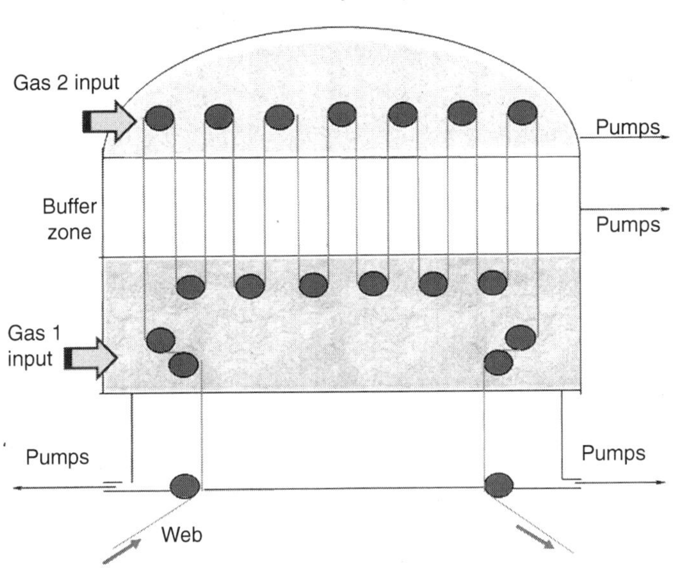

Figure 7.34 A schematic of an ALD process using winding geometry to build up the coating thickness.

has been used for barrier coatings during the development. As with the other design there is still more development work to be done before this becomes a major competitor for the production of barrier coatings. The existing design has what could be seen by different users as either a disadvantage or an advantage. As the gas sticks to all surfaces it means that the gas sticks to both the front and back surface of the film and so the coating grows on both surfaces. This is fine if you want a coating on both surfaces and the barrier looks to be good but part of the improvement is because any pinholes are offset and there is a more tortuous diffusion path. It is possible to engineer the system to limit the back surface deposition but care is needed to make sure that no film damage is done in using static masking.

This is certainly a deposition process for the future and looks to have the potential to reduce the transparent barrier costs because of reduced system capital costs. As with chemical vapour deposition the economics of this does also depend on having the appropriate gaseous precursors and at a reasonable cost. This also means that the precursor gases need to be in the low hazard category otherwise the gas handling costs or pumping costs increase significantly.

7.7 Other Deposition Processes

Currently there is research and development being done to try to increase the deposition efficiency of the aluminium metallizing process. Part of the problem with the existing sources is that is in some systems more than 50% of the evaporant does not coat the substrate but is collected by the various system shields. Sources have been developed that can have material collection efficiencies of >95% (192) and similar principles are being used to develop sources suitable for use with polymer substrates. Depending on the design some of the sources have been named as jet vapour sources. These sources, once they become available, will help reduce the deposition cost as for the same energy almost twice the polymer area can be coated. Also these sources allow the heat load to be distributed more evenly over a larger substrate area and so the winding speed can be increased and this too can lead to cost reductions.

As described in chapter 5.5 it is possible to deposit polymer coatings inside the vacuum system. This can be done either by the flash evaporation of a monomer inside a closed volume inside the vacuum chamber. Direct evaporation of polymers is also being done from slot sources that also give a good uniformity across the web width. These sources are usually based on radiant heated evaporation source design. A key feature of these designs is a very high level of temperature uniformity over the whole of the source which is essential to control the evaporation rate.

References

1. Mount E.M. III & Bishop C.A. Eds *'Metallizing Technical Reference'* 4th Edn. Pub. Association of Industrial Metallizers, Coaters & Laminators (AIMCAL) 2007.
2. d'Ouville T. et al. 'The effects of boat & wire parameters on boat life & coating in vacuum metallization of an OPP web.' Proc. 38th Ann. Tech. Conf. SVC 1995 pp 125–131.
3. Baxter I. 'Advance resistance deposition technology for productive roll coating.' Proc. 36th Ann. Tech. Conf. SVC 1993 pp 197–204.
4. Ruisinger B. et al. 'Evaporation boats – investigations on the electrical properties.' Proc. 35th Ann. Tech. Conf. SVC 1992 pp 84–87.
5. Ruisinger B. & Mossner B. 'Evaporation boats – Properties, requirements, handling & future requirements.' Proc. 34th Ann. Tech. Conf. SVC 1991 pp 335–338.

6. Fletcher D.A. 'Evaporation boats – latest developments improve metallizing performance & economics.' Proc. 31st Ann. Tech. Conf. SVC 1988 pp 267–273.
7. Zoeliner H. 'Resistance heated ceramic evaporation boats.' Proc. 27th Ann. Tech. Conf. SVC 1984 pp 1–4.
8. Mitchell R. et al. 'Influencing barrier properties by optimising aluminium wire & evaporation boats.' Proc. 10th Vac. Web Coat. Conf. 1996 pp 134–145.
9. Mount E.M. III & Bishop C.A. Eds *'Metallizing Technical Reference'* 4th Edn. Pub. Association of Industrial Metallizers, Coaters & Laminators (AIMCAL) 2007.
10. Archibald P. & Parent E. 'Source evaporant systems for thermal evaporation' *Solid State Technology* July 1976 pp 32–40.
11. Parent E.D. 'Power requirements of resistance heated intermetallic evaporation sources' *J. Vac. Sci. & Tech. Vol II,* No. 4 1974, pp 820–823.
12. Gibson C.D. & Kohnken K.H. 'Intermetallic evaporator boats.' Proc. 3rd Vac. Web. Coat. Conf. 1989 pp 121–130.
13. Nuernberger M. 'Changes to an IMC Evaporator During Aluminium Evaporation' Proc. AIMCAL Fall Tech. Conf. 2007, 21st Internat. Vacuum Web Coating Conf.
14. Schmoll U. 'Impact of Corrosion Resistance on the Performance of Evaporator Boats' Proc. AIMCAL Fall Tech. Conf. 2009, 23rd Internat. Vacuum Web Coating Conf.
15. Susaki J. 'Precise Detection of Evaporation Boat Breakdown' Proc. AIMCAL Fall Tech. Conf. 2007, 21st Internat. Vacuum Web Coating Conf.
16. Meinel J. et al. 'Dynamics & efficiency aspects in the evaporator zone of vacuum metallizers.' Proc. 36th Ann.Tech.Conf. SVC 1993 pp 191–196.
17. Pasqui A. et al. 'High rate deposition on new high speed aluminium roll coaters.' Proc. 36th Ann.Tech.Conf. SVC 1993 pp 185–189.
18. Swisher R.L. 'Distribution patterns of coatings from small sources onto webs moving over drums.' Proc. 7th Intnl. Vac.Web Coat. Conf. 1993 pp 64–69.
19. Casey F & Ellis G H. 'Deposition uniformity & visual appearance in high speed web coaters.' Proc. 10th Intnl. Vac. Web Coat. Conf. 1996 pp 123–132.
20. Zelisse J.K. 'Accurate modelling of eddy current sensors for square resistance & thickness measurements' Proc. 5th Intnl. Vac. Web Coating Conf. 1991 pp 135–143.
21. Duesbury P. 'A heuristic approach to temperature compensation in non-contact resistance meters for vacuum roll coaters.' Proc. 36th Ann. Tech. Conf. SVC 1993 pp 232–235.

22. Brandenberg J. et al. 'Experiences of operating a closed-loop feedback system on a vacuum web coater.' Proc. 35th Ann.Tech.Conf. SVC 1992 pp 94–99.
23. Scarr J.M. & Zelisse J.K. 'A new topology for thickness monitoring eddy current sensors' Proc. 36th Ann. Tech. Conf. SVC 1993 pp 228–231.
24. Felts J. 'Transparent gas barrier technologies' Proc 33rd Ann. Tech. Conf. SVC 1990, pp 184–193.
25. Krug T. G. 'Transparent barriers for food packaging' Proc. 33rd Ann. Tech. Conf. SVC 1990, pp 163–169.
26. Chahroudi D. 'Transparent Glass Barrier Coatings for Flexible Film Packaging' Proc. 34th Ann Tech Conf SVC 1991, pp 130–134.
27. Bowler J.F. 'Demand & supply for high barrier films.' Proc. 5th Vacuum Web Coating Conf. 1991 pp 72–78.
28. Gavitt I.F. 'Snack food packaging: how much is enough.' Proc. 37th Ann.Tech.Conf. SVC 1994 pp 127–132.
29. Lohwasser w. et al. 'Electron-Beam Oxide Coating on Plastic Films for Packaging, Development, Production and Application' Proc. 38th Ann Tech Conf SVC 1995, pp 40–46.
30. Fayet P. 'Commercialisation of plasma deposited barrier coatings for liquid food packaging.' Proc. 38th Ann.Tech.Conf. SVC 1995 pp 15–16.
31. Finson E. & Hill R. 'Glass-coated packaging films ready for commercialization' Pack. Technol. & Engin. June/July 1995, pp 36–43.
32. Chatham H. 'Oxygen diffusion barrier properties of transparent oxide coatings on polymeric substrates.' *Surf. Coat. Technol.* Vol. 78 1996 pp 1–9.
33. Leterrier Y. 'Durability of nanosized oxygen-barrier coatings on polymers.' Prog. in *Matls. Sci.* Vol. 48, 2003 pp 1–55.
34. Kelly R.S.A. 'Development of Clear Barrier Films in Europe' Proc. 36th Ann. Tech. Conf. 1993 pp 312–317.
35. Kelly R.S.A. 'Development of Aluminium Oxide Clear Barrier Films' Proc. 37th Ann. Tech. Conf. 1994 pp 144–149.
36. Kelly R.S.A. 'Transparent barrier coatings for films' Proc. AWA Conf. Metallized papers & Films – Europe, Feb. 2002 paper IX.
37. Web barrier packaging material European Patent Application EP0437946 1990.
38. Kobayashi T. et al. 'Reactive vacuum vapour deposition of aluminium oxide thin films by an air-to-air metallizer' *J. Vac. Sci. Technol. A* Volume 24, Issue 4, pp. 939-945 (July 2006).
39. Schiller S. et al. 'How to produce Al_2O_3 coatings on plastic films' 7th Internat. Conf on Vac. Web Coating. 1993 pp 194–219.
40. Schiller S. et al. 'Al_2O_3 layers on plastic films – ways to get high deposition rates' 8th Internat. Conf on Vac. Web Coating. 1994 pp 173–190 .
41. Schiller N. et al. 'Innovative clear barrier technology for the packaging industry' Proc. AIMCAL Fall Tech Conf 2008.

42. Schiller N. et al. 'Transparent Barrier Coatings on Polymer Films' Proc. AIMCAL Fall Tech Conf 2009.
43. Maddocks J. et al. 'Packaging barrier films deposited on PET by PECVD using new high density plasma source.' Proc. 47th Ann.Tech. Conf. SVC 2004 pp 672–673.
44. Rochat G et al. 'Cohesive strength and oxygen permeability of PECVD-based coatings on semicrystalline polymers.' Proc. 46th Ann. Tech.Conf. SVC 2003 pp 50–55.
45. Komada M. et al. 'Novel transparent gas barrier film prepared by PECVD method.' Proc. 43rd Ann.Tech.Conf. SVC 2000 pp 352–356.
46. Sacher E. et al. 'Moisture barrier properties of plasma-polymerized hexamethyldisiloxane.' *J. Appl. Polymer Sci: Appl. Polymer Symp.* 38, 1984 pp 163–171.
47. Affinito J. et al. 'A new hybrid deposition process, combining PML & PECVD, for high rate plasma polymerisation of low vapor pressure, and solid, monomer precursors.' Proc. 42th Ann.Tech.Conf. SVC 1999 pp 102–107.
48. Weston G.F. 'Materials for ultrahigh vacuum.' *Vacuum* Vol. 25, No.11/12 pp 469–484.
49. Smith A.W. et al. 'PECVD of SiOx barrier films.' Proc. 45th Ann.Tech. Conf. SVC 2002 pp 525–529.
50. Matteucci J.S. 'Electron beam evaporation and D.C. magnetron sputtering in roll coating.' Proc. 30th Ann. Tech. Conf. SVC 1987 pp 91–104.
51. Schiller S. et al. 'Electron beam coating' in '*Surfacing Technologies Handbook*' Ed. Sudarshan T.S. Pub. Marcel Dekker 1988.
52. Pierce J.R. '*Theory of design of electronic beams.*' Pub. Van Nostrand Inc. 1954.
53. Schiller S. & Neumann. 'High rate electron beam evaporation for roll coating.' Proc. Conf. EB Melting & Refining 1984.
54. Schiller S. et al. 'High-rate electron beam evaporation.' *Thin Solid Films* Vol. 110, 1983 pp 149–164.
55. Schiller S. & Neumann M. 'Application of EB line evaporators for web coating.' Proc. 1st Internat. Vacuum Web Coating Conf. 1987 pp 113–128.
56. Schiller S. et al. 'Possibilities of silicon oxide deposition in vacuum web coating' Proc. 3rd Internat. Vacuum Web Coating Conf. 1989 pp 24–49.
57. Schiller S. et al. 'Capabilities of electron beam evaporation for the deposition of compounds.' Proc. 4th Internat. Vac. Web Coat. Conf. 1990 pp 138–158.
58. Schiller S. et al. 'Advances in electron beam deposition of SiOx films' Proc. 5th Internat. Vac. Web Coat. Conf. 1991 pp 19–34.
59. Heinss J-P. et al. 'Enhanced surface properties of metallic sheets by high-rate electron beam evaporation.' Proc. 40th Ann. Tech. Conf. SVC 1997 pp 157–162.

60. Schiller S. et al. 'Plasma-activated high-rate deposition of oxides on plastic films.' Proc. 37th Ann. Tech. Conf. SVC 1994 pp 203–210.
61. Hartwig E. & Krug T. 'Food packaging films utilizing non-metallic & glass-like barriers.' Proc. Speciality Plastics Conf. Zurich 1991.
62. Krug T.G. 'Transparent barriers for food packaging' 33rd Annual Technical Conference Proceedings SVC 1990 pp 163–169.
63. Scateni G. et al. 'Transparent barrier coatings: test-runs on industrial equipment.' Proc. 5th Vacuum Web Coating Conf. 1991 pp 104–113.
64. Schiller S. et al. 'Plasma-Activated High-Rate Deposition of Oxides on Plastic Films' 37th Annual Technical Conference Proceedings SVC 1994 pp 203–210.
65. Taguchi T et al. 'Air-to-air metallizer: design & operational data.' Proc. 35th Ann. Tech. Conf. SVC 1992 pp 135–140.
66. Yadin E. & Andreev Y. 'Zinc & magnesium vapor generators in a steel strip coating system.' Proc. 42nd Ann. Tech Conf. SVC 1999 pp 39–42.
67. Yadin E. 'Deposition of coatings or free foils of sublimating metals.' Proc. 40th Ann. Tech Conf. SVC 1997 pp 390–393.
68. Tada I. et al. 'Comparison of evaporation sources for vacuum web coaters.' Proc. 32nd Ann. Tech Conf. SVC 1989 pp 131–149.
69. Miyabayashi T et al. 'Recent advances on the world largest 122 inch vacuum web coater aimed at cast polypropylene (CPP) films. Proc. 3rd Internat. Web Coating Conf. 1989 pp 2–23.
70. Guentherschultze A. 'Cathodic sputtering – as analysis of the physical processes.' *Vacuum* Vol. 3, No. 4, 1953 pp 360–374.
71. Westwood W.D. 'Glow discharge sputtering.' Progress in Surf. Sci. Vol.7, 1978 pp 71–111.
72. Wehner G.K. & Anderson G.S. 'The nature of physical sputtering.' Chptr 3 Maissel L. 'Application of sputtering to the deposition of films.' Chptr 4. '*Handbook of thin film technology*.' Maissel L.I. & Glang R. Pub. McGraw-Hill 1983 (reissue) ISBN 0 07 039742 2.
73. Greene J.E. 'Epitaxial crystal growth by sputter deposition: applications to semiconductors. Part 1.' *CRC Crit. Rev. in Solid State & Matls. Sci.* 1984 Vol. 11, No. 1, pp 47–97.
74. Ectertova L. '*Physics of thin films*.' Pub. Plenum 2nd Edn 1986 ISBN 0 306 41798 7.
75. Adachi R. & Takeshita K. 'Magnetron sputtering with additional ionization effect by electron beam.' *J. Vac. Sci & Tech.* Vol.20. No.1, 1982 pp 98–99.
76. Thornton J.A. & Penfold A.S. 'Cylindrical magnetron sputtering.' Chptr II-2 Vossen J.L. & Kern W. '*Thin film processes*.' Academic Press Inc. 1978 ISBN 0 12 728250 5.
77. Chapin J.S. 'The planar magnetron.' *Research & Development* Vol. 25 No. 1 1974 pp 37–40.

78. US 4,166,018 patent 'Sputtering process & apparatus.' Chapin J.S. 1979.
79. US 3,884,793 Patent 'Electrode type glow discharge apparatus,' Penfold A.S. & Thornton J.A. 1975.
80. Thornton J.A. 'Magnetron sputtering: basic physics & application to cylindrical magnetrons.' *J. Vac. Sci & Tech.* Vol.15. No.2, 1978 pp 171–177.
81. US 4,162,954 Patent. 'Planar magnetron sputtering device.' Morrison C.F. 1979.
82. US 4,180,450 Patent. 'Planar magnetron sputtering device.' Morrison C.F. 1979.
83. US 4,198,283 Patent. 'Magnetron sputtering target & cathode assembly.' Class W.H. et al 1980.
84. US 4,282,083 Patent. 'Penning sputter source.' Kertesz G & Vago G. 1981.
85. US 4,444,643 Patent. 'Planar magnetron sputtering device.' Garrett C.B. 1984.
86. US 4,457,825 Patent. 'Sputter target for use in a sputter coating source.' Lamont Jr. L.T. 1984.
87. US 4,426,264 Patent. 'Method & means for controlling sputtering apparatus.' Munz W-D & Wolf H. 1984.
88. Spencer A.G. et al. 'The design & performance of planar magnetron sputtering cathodes.' Vacuum 1987 Vol. 37 Nos. 3/4, pp 363–366.
89. US 4,746,417 Patent. 'Magnetron sputtering cathode for vacuum coating apparatus.' Ferenbach D. & Muller J. 1988.
90. US 4,761,218 Patent. 'Sputter coating source having plural target rings.' Boys. D.R. 1988.
91. Window B. & Savvides N. 'Charged particle fluxes from planar magnetron sputtering sources.' *J. Vac. Sci. Technol. A* Vol.4, No.2, 1986 pp 196–202.
92. Window B. & Savvides N. 'Unbalanced dc magnetrons as sources of high ion fluxes.' *J. Vac. Sci. Technol. A* Vol. 4, No. 3, 1986 pp 453–465.
93. Savvides N. & Window B. 'Unbalanced magnetron ion-assisted deposition & property modification.' *J. Vac. Sci. Technol. A* Vol. 4, No.3, 1986 pp 504–508.
94. Munz W-D. 'Current industrial practices. The unbalanced magnetron: current status of development.' *Surf. & Coat. Technol.* Vol. 48, 1991 pp 81–94.
95. Kelly P.J. et al. 'Novel materials by unbalanced magnetron sputtering.' Materials World March 1993 pp 161–165.
96. US 5,556,519. 'Magnetron sputter ion plating.' Teer D.G. Pub. 1996.
97. Kelly P.J. & Arnell R.D. 'Magnetron sputtering: a review of recent developments & applications.' *Vacuum* Vol. 56, 2000 pp 159–172.
98. Schatz D.S. 'The MDX™ as a strategic tool in reducing arcing.' Applied Energy Industries Inc. Application notes. 1985.

99. Sellers J. 'Asymmetric bipolar pulsed DC.' ENI® Tech. Note 1985.
100. Grove T.C. 'Arcing problems encountered in sputter deposition of aluminium.' Proc. 29th Ann. Tech. Conf. SVC 1986 pp 214–223.
101. Scholl R.A. 'Process improvements for sputtering carbon & other difficult materials using combined AC & DC process power.' Proc. 35th Ann. Tech. Conf. SVC 1992 pp 391–394.
102. Anderson G.L. 'A new technique for arc control in DC sputtering.' Proc. 35th. Ann. Tech Conf. SVC 1992 pp 325–329.
103. Scholl R.A.'A new method of handling arcs & reducing particulates in dc plasma processing.' Proc. 36th. Ann. Tech Conf. SVC 1993 pp 405–408.
104. US 5,241,152 patent. 'Circuit for detecting & diverting an electrical arc in a glow discharge apparatus.' Anderson G.L. et al 1993.
104. Harpold J. & Scholl R.A. 'How Advanced Energy® MDX™ products manage arcs.' Applied Energy Industries Inc. Application notes. 1993.
105. Scholl R.A. 'Advances in arc handling in reactive & other difficult processes.' Proc. 37th. Ann. Tech Conf. SVC 1994 pp 312–316.
106. Sellers J.C. 'Asymmetric bipolar pulse DC - an enabling technology for reactive PVD.' Proc. 39th. Ann. Tech Conf. SVC 1996 pp 123–126.
107. Sellers J. 'Using the RPG generator.' ENI® Tech. Note 1996.
108. Brauer G. et al. 'Mid frequency sputtering–A novel tool for large area coating. Surfaces & Coatings Vols. 94-95, 1997 pp 258–262.
109. Schneider J.M. & Sproul D.M. 'Reactive pulsed DC magnetron sputtering & control.' In *'Handbook of thin film process technology.'* Eds. Glocker D.A. & Shah S.I. Pub. Inst. of Phys. Bristol & Philapelphia. 1998 Section A5.1.
110. Este G.O. & Westwood W.D. 'AC & RF reactive sputtering.' in *'Handbook of thin film process technology.'* Eds. Glocker D.A. & Shah S.I. Pub. Inst. of Phys. 1998 Section A5.2.
111. Scholl R.A. 'Power supplies for pulsed power technologies: state-of-the-art & outlook.' Applied Energy Industries Inc. Application notes. 1999.
112. Koski K. et al. 'Surface defects & arc generation in reactive sputtering of aluminium oxide thin films.' *Surf. Coat. Technol.* Vol. 115, 1999 pp 163.
113. Scholl R.A. et al. 'Problems & solutions in reactive sputtering of dielectrics.' Proc. 13th Internat. Vac. Web Coat. Conf. 1999 pp 153–164.
114. De. Bosscher A. et al. 'Global solution for reactive magnetron sputtering. Proc. 3rd Internat. Conf. Coatings on Glass 2000 pp 59–76.
115. Rettich T. & Wiedemuth P. 'MF, DC & pulsed DC in the use of large area coating applications on glass. Proc. 3rd Internat. Conf. Coatings on Glass 2000 pp 97–105.
116. Blondeel A. & Bosscher W.De. 'Arc handling in Reactive DC magnetron sputter deposition.' Proc. 44th. Ann. Tech Conf. SVC 2001 pp 240–245.

117. Milde F. et al. 'Experience with high power DC supplies with fast arc suppression in large area coating.' Proc. 44th. Ann. Tech Conf. SVC 2001 pp 375–381.
118. Segers A. et al. 'Arc discharges in the reactive sputtering of electrical insulating dielectrics.' Proc. 45th. Ann. Tech Conf. SVC 2002 pp 30–35.
119. Christie D.J. et al. 'A novel pulsed supply with arc handling & leading edge control as enabling technology for high powered pulsed magnetron sputtering HPPMS.' Proc. 47th. Ann. Tech Conf. SVC 2004 pp 113–118.
120. Rettich T. et al. 'Arc management in DC & MF generators for large area coating systems.' Proc. 47th. Ann. Tech Conf. SVC 2004 pp 237–240.
121. US 4,478,702 patent. 'Anode for magnetic sputtering apparatus.' Gillery F.H. & Criss R.C. 1984.
122. US 4,744,880 patent. 'Anode for magnetic sputtering of gradient films.' Gillery F.H. et al 1988.
123. Sieck P.A. et al. 'Anode structure for magnetron sputtering systems.' US 5,683,558 patent 1997.
124. Sieck P.A. et al. Magnetron sputtering methods and apparatus. Patent EP 0,674,337 A1 1995.
125. Belkind A. et al. 'Reactive sputtering using a dual-anode magnetron system.' Proc. 44th Ann. Tech. Conf. SVC 2001 pp 130–135.
126. Scholl R. et al. 'Anode problems in pulsed power reactive sputtering of dielectrics.' Proc. 42nd Ann. Tech. Conf. SVC 1999 pp 169–175.
127. Glocker D.A. 'An estimate of potentials developed on coated anodes during pulsed dc reactive sputtering.' Proc. 43rd Ann.Tech.Conf.SVC 2000 pp 87–90.
128. Sieck, P. 'Active Control of Anode Current Distribution for D.C. Reactive Sputtering of SiO_2 and Si_3N_4.' *Surf. Coat. Technol.* (Switzerland). Vol. 68/69, pp.794-798. 1 Dec.1994.
129. Belkind A & Jansen F. 'Anode effects in magnetron sputtering.' *Surf. Coat. Technol.* Vol. 99, 1998 pp 52–59.
130. Scholl R. et al. 'Problems & solutions in reactive sputtering of dielectrics.' Proc. 13th Vacuum Web Coating. Conf. 1999 pp 153–164.
131. Brauer G. et al. 'New approaches for reactive sputtering of dielectric materials on large scale substrates.' *J. Non-Crystalline Solids* Vol.218, 1997 pp 19–24.
132. US 4,572,776 Patent. 'Magnetron cathode for sputtering ferromagnetic targets.' Aichert H. et al 1986.
133. US 4,519,885 Patent. 'Method & apparatus for changing sputtering targets in a magnetron sputtering system.' Innis D.T. 1985.
134. EP 0,134,559 Patent. 'Cathodic sputtering apparatus.' McKelvey H.E. 1985.
135. Wright M & Beardow. T. 'Design advances & applications of the rotatable cylindrical magnetron.' *J. Vac. Sci. Technol. A,* Vol. 2, No. 3, Pt. 1, 1986 pp 388–392.

136. CA 1,239,115 Patent. 'Sputtering apparatus & method.' Kovilvila R. 1988.
137. EP 0,045,822 Patent. 'Cylindrical magnetron sputtering cathode, as well as sputtering apparatus provided with such cathodes.' Zega B. 1982.
138. Hill R.J. & Nadel S.J. 'Coated glass: applications & markets.' Pub. BOC Coating Technol. Fairfield, CA. 1999 ISBN 0-914289-01-2.
139. Spencer A.G. et al. 'Pressure stability in reactive magnetron sputtering.' *Thin Solid Films* Vol. 138, 1988 pp 141–149.
140. Spencer A.G. & Howson R.P. 'Dynamic control of reactive magnetron sputtering: A theoretical analysis.' *Thin Solid Films* Vol.138, 1988 pp 129–136.
141. Spencer A.G. et al. 'Design & use of a vacuum system for high rate reactive sputtering of TiO_2/TiN/ TiO_2 solar control films.' *Solar Energy Materials* Vol. 18, 1988 pp 87–95.
142. Spencer A.G. et al. 'Pressure stability in reactive magnetron sputtering,' Proc. IPAT 1987 pp 1–6 Pub. CEP Consultants Ltd.
143. Bishop C.A. '*The deposition of coatings onto polymer substrates by planar magnetron sputtering.*' PhD Thesis 1986 Loughborough University. UK.
144. Penfold A.S. The influence of pump size on D.C. reactive sputtering systems. Proc. 29th Ann. Tech. Conf. SVC 1986 pp 381–403.
145. Berg S. et al. 'Modelling of reactive sputtering of compound materials.' *J. Vac. Sci. Technol. A,* Vol. 5, No. 2, 1987 pp 202–207.
146. Larsson T. et al. 'A physical model for eliminating instabilities in reactive sputtering.' *J. Vac. Sci. Technol. A,* Vol. 6, No. 3, 1988 pp 1832–1836.
147. Berg S. et al. 'Predicting thin-film stoichiometry in reactive sputtering.' *J. Appl. Phys.* Vol. 63, No. 3, 1988 pp 887–891.
148. Carlsson P et al. 'Reactive sputtering using two reactive gases, experiments and computer modelling.' *J. Vac. Sci. Technol. A,* Vol. 11, Issue 4, pp 1534–1539.
149. Serikawa T. & Okamoto A. 'Effect of N_2—Ar mixing on the reactive sputtering characteristics of silicon.' *Thin Solid Films* Vol. 101, Issue 1 March 1983 pp 1–6.
150. Kadlec S. et al. 'Hysteresis effect in reactive sputtering: a problem of system stability.' *J. Phys. D* Vol. 19, No 9 Sept. 1986 pp L187.
151. Okamoto A. & Serikawa T. 'Reactive sputtering characteristics of silicon in an Ar—N2 mixture.' *Thin Solid Films* Vol. 137, Issue 1, March 1986 pp 143-151.
152. Sproul W.D. 'High rate reactive sputtering process control.' *Surface & Coatings Technol.* Vol. 33, 1987 pp 73–81.
153. Bellido-Gonzalez V. et al. 'Flexible reactive gas sputtering process control.' Proc. 47th Ann. Tech. Conf. SVC 2004 pp 44–48.
154. Ershov A. & Pekker L. 'Model of DC magnetron reactive sputtering of the silicon target in Ar-0_2 gas mixtures. Proc. 39th Ann. Tech. Conf. SVC 1996 pp 279–284.

155. Schiller S. et al. 'Different properties of reactively DC sputtered chromium-silicon-oxide films with variation of target-to-substrate coupling.' Proc. 7th Internat. Conf. Ion & Plasma Assisted Techniques. (IPAT) Pub. CEP Consultants. 1989 pp 124–129.
156. Maniv S. et al. 'High rate deposition of transparent conducting films by modified reactive planar magnetron sputtering of Cd_2Sn alloy.' *J. Vac. Sci. Technol. A*, Vol. 18, Issue 2, 1981 pp 195–198.
157. Belkind A. & Wolfe J. 'Enhancement of reactive sputtering rate of TiO2 using a planar and dual rotatable cylindrical magnetrons.' *Thin Solid Films* Vol. 248, Issue 2, Aug. 1984 pp 163–165.
158. Westwood W.D. 'Reactive sputtering: introduction and general discussion.' *'Handbook of thin film process technology.'* Eds. Glocker D.A. & Shah S.I. Pub. Inst. of Phys. 1998 Section A5.0.
159. Howson R.P. & Spencer A.G. 'Self-pumping in high rate reactive sputtering.' Proc. 32nd Ann. Tech. Conf. SVC 1989 pp 40–47.
160. 'Advanced vacuum web coating trends reviewed.' *Paper, film & foil converting*. Apl 1989 pp 70–72.
161. Buschbeck W. et al. 'Vacuum design for mixed-mode sputtering roll coaters.' Proc. 30th Ann. Tech. Conf. SVC 1987 pp 127–134.
162. Schiller S. et al. 'Reactive D.C.high-rate sputtering as production technology.' Surfaces & Coat. Technol. Vol. 33, 1987 pp 405–423.
163. Howson R.P. et al. 'The importance of gas flow in reactive sputtering.' Proc. 8th Internat. Conf. Ion & Plasma Assisted Techniques. (IPAT) Pub. CEP Consultants. 1991 pp 340–345.
164. Schiller S. et al. 'Progress in the application of the plasma emission monitor in web coating.' Proc. 2nd Internat. Vacuum Web Coat. Conf. 1988 pp 124–138.
165. Hmiel A.F. 'partial pressure control of reactively sputtered titanium nitride.' *J. Vac. Sci. Technol.* A, Vol. 3, No. 3, 1985 pp 592–595.
166. Schiller S. et al. 'Fully automated control of reactive sputtering processes in web coaters by integrated PEM stabilisation.' Proc. 1st Intnatl. Vacuum Web Coating Conf. 1987 pp 54–61.
167. Greene J.E. 'Optical spectroscopy for diagnostics & process control during glow discharge etching & sputter deposition. *J. Vac. Sci. Tech.* 1978 Vol. 15, No. 5, pp 1718–1729.
168. Strumpfel J. et al. 'Production of optical multilayers on webs for Ar application by means of reactive dual magnetron sputtering.' Proc. 11th Vacuum Web Coating Conf. 1997 pp 279–289.
169. Fraval R.H. & Angus R. 'A novel optical spectrum analyser for plasma process monitoring.' Proc. IPAT Workshop 'Semiconductor technology–Thin film production & properties.' June 1986 Pub. CEP Consultants Ltd.
170. Cilia M. et al. 'Optical emission spectroscopy study of the radio-frequency magnetron discharge used for the fabrication of X-ray

multilayer mirror.' *Thin Solid Films* Vol. 312, Nos. 1/2, 1998 pp 320–326.
171. Schiller S. et al. 'Application of plasma emission spectroscopy for process control in reactive D.C. high-rate plasmatron sputtering.' Proc. Internat. Conf. Met. Coat. Apl 1983 pub. in *Thin Solid Films* Vol. 108, 1983.
172. Enjouji K. et al. 'The analysis & automatic control of a reactive d.c. magnetron sputtering process.' *Thin Solid Films* Vol. 108 1983 pp 1–7.
173. Rossnagel S.M. & Saenger K.L. 'Optical emission in magnetrons: nonlinear aspects.' *J. Vac. Sci. Technol. A* Vol. 7, No. 3, 1989 pp 968–971.
174. Theil J.A. 'Gas distribution through injection manifolds in vacuum systems.' *J. Vac. Sci. Technol. A*, Vol. 13, No. 2, 1995 pp 442.
175. Bartzsch H. & Frach P. 'Modelling the stability of reactive sputtering processes.' Proc. 7th Internat. Conf. on Plasma Surf. Eng. 2000.
176. Milde T. et al. 'Gas inlet systems for large area linear magnetron sputtering sources.' Proc. 44th Ann. Tech. Conf. SVC 2001 pp 204–209.
177. Spencer A.G. '*High rate reactive magnetron sputtering*.' PhD Thesis 1989 Loughborough University.
178. Butler H.S. & Kino G.S. 'Plasma sheath formation by radio-frequency fields.' *The Physics of Fluids* Vol. 6, No. 9, 1963 pp 1346–1355.
179. Tsui R.T.C. 'Calculation of ion bombarding energy & its distribution in rf sputtering.' *Phys. Rev.* Vol. 168, No. 1, 1968 pp 107–113.
180. Jackson G.N. 'R.F. Sputtering: A review.' *Thin Solid Films* Vol. 5, 1970 pp 209–246.
181. Lau S.S. et al. 'Temperature rise during film deposition by rf & dc sputtering.' *J. Vac. Sci. Technol.* Vol. 9, No. 4, 1972 pp 1196–1202.
182. Norstrom H. 'Experimental & design information for calculating impedance matching networks for use in rf sputtering & plasma chemistry.' *Vacuum* Vol. 29, No. 10, 1979 pp 341–350.
183. Gill M.D. 'Sustaining mechanisms in rf plasmas.' *Vacuum* Vol. 34, Nos. 3/4, 1984 pp 357–364.
184. Kohler K. et al. 'Plasma potentials on 13.56-MHz rf argon glow discharges in a planar system.' *J. Appl. Phys.* Vol. 57, No. 1, 1985 pp 59–66.
185. Steinbruchel. Ch. et al. 'Diagnostics of low-pressure oxygen rf plasmas & the mechanism for polymer etching: A comparison of reactive sputter etching & magnetron sputter etching.' *IEEE Trans. on Plasma Sci.* Vol. PS-14, No. 2, 1986 pp 137–144.
186. Savas S.E. et al. 'Dummy load technique for power efficiency estimation in rf discharges.' *Rev. Sci. Instrum.* Vol. 57, No. 7, 1986 pp 1248–1250.

187. Brouk V. & Heckman R. 'Stabilizing RF generator & plasma interactions.' Proc. 47th. Ann. Tech Conf. SVC 2004 pp 49–54.
188. Sneck S. 'High Capacity Atomic Layer Deposition for Industrial Coating Applications' Proc. 50th Ann. Tech. Conf. SVC 2007 pp 36–38.
189. Sneck S. 'Atomic Layer Deposition in Mass Production of Optical Coatings' Proc. 51st Ann. Tech. Conf. SVC 2008 pp 413–416.
190. Levy D. 'Spatial atomic layer deposition: High quality films on continuous substrates' Proc. AIMCAL Fall Tech Conf. 23rd Internat. Vac.Web Coating Conf. 2009.
191. Barrow W.A. 'Roll-to-roll ALD deposition of Al_2O_3 barrier layers on PET' Proc. AIMCAL Fall Tech Conf. 23rd Internat. Vac.Web Coating Conf. 2009.
192. Franciscus J.A. et al. 'Method and device for coating a substrate' United States Patent 7,323,229 B2 Jan 29, 2008.

8

Summary

This summary is really aimed at providing a reminder of the most critical aspects that impinge on the barrier performance of vacuum deposited coatings.

8.1 Cleanliness

As defects reduce the performance of any of the barrier coatings and the first few defects are the most damaging it is important to review the whole process and make sure all aspects remain as clean as possible.

It is expected that the substrate will be as clean as possible to start with and wherever necessary the substrate will be cleaned before use.

Where the roll is handled, whether it is using a winding system run in atmosphere to deposit a subbing, smoothing or planarising layer before the vacuum coating process, or the vacuum coater winding system itself all must be kept as clean as possible. If a polymer planarising layer is coated it is worth considering using a cleanroom or clean positive pressure filtered air hoods to make sure the coating is not contaminated after coating.

The vacuum system will get dirty because of the nature of the vacuum deposition process and there is the need to clean any masking shields. If this is not cleaned regularly and cleaned well there will be dust stirred up during the system pump down and the system venting process. This stirred up dust will reach the winding system and can lead to coating defects including pinholes.

This attention to cleanliness does not stop after the vacuum deposition process as the coating will still be fragile and any contamination downstream of the vacuum coating process can lead to scratches and coating cracking.

8.2 Substrates

Ideally the substrates will have all the mechanical and optical attributes required as well as being thermally stable, smooth and flat with no surface contamination and high surface energy. As substrates are not normally available with all of these attributes it will be necessary to manage what is available.

The substrates should be cleaned to minimize the particulate contaminants. If costs allow it would be preferred for the remaining particulate and exuded contaminants to be covered up producing the smoothest and most perfect surface possible.

Again if costs permit, using a heat stabilized grade will allow a higher deposition temperature with minimum dimensional effect on the substrate.

The substrates should be plasma treated to optimise the adhesion but this requires care as it is possible to over treat the surface which can reduce the adhesion and also add surface roughness that will lead to coating defects and cracking and loss of barrier performance.

Substrates should be of good quality as in they should be of good profile with a high quality of slit edge and wound with an even tension. Too much tension will damage the roll and too little tension can lead to telescoping in the vacuum system. Poor profile or slitting will give rise to high spots that will take more tension and may lead to winding problems. The vacuum winding system should also be of the highest quality as once the coating has been deposited it can be fragile and so the re-wind roll does not want to slip though poor tension control or a too high applied tension.

8.3 Coatings

The target is to deposit the smoothest most dense but stress free coating possible. Cost can enter into this equation. Resistance heated source evaporation is the simplest and cheapest deposition process but the coatings produced will not be as dense as if the same materials were to be magnetron sputtered onto the substrate. However sputtering has a deposition rate of the order of one thousand times slower than that of evaporation. Electron beam evaporation provides an intermediate solution as there is a slightly higher deposition energy whilst retaining the high deposition rate. This can be further enhanced using an additional plasma source situated between the source and substrate. In this way the coatings can be densified and the high deposition rate maintained. The danger is that the coatings whilst dense will also have a higher stress which can also lead to coating failure by cracking or delamination.

Within the deposition system process stability also helps minimize defects irrespective of which deposition process is being used. The highest possible stability of the molten pool in the resistance heated aluminium evaporation boats minimises the production of spits. In magnetron sputtering keeping the target surface clean of backscattered material helps minimise micro-arcing and coating damage. In any of the plasma enhanced chemical vapour deposition processes maintaining the plasma uniformity by keeping the anode area active helps maintain a constant deposition rate and any plasma intensity variations.

Although this attention to detail in keeping the system clean and stabilising the deposition process as much as possible it will still be hard to eliminate all defects.

System design can help as it may be possible to arrange the winding system to have no front surface contact rolls after the deposition zone. This will help prevent any post deposition damage into the freshly deposited coatings. Aluminium is particularly soft until the native surface oxide layer has build up, which usually takes one or two days. In extremely sensitive systems it has been known to include a cleanroom grade paper interlayer between the layers in the re-wind roll. The aim of this interlayer is that it is softer than the coating and so prevents any surface damage to the coating.

8.4 Over Coatings

In the same way that paper interleaving can help prevent any coating damage so too can using a polymer over coating. Not only protecting the inorganic coating but also the organic coating can flow and fill in any coating defects which will reduce the diffusion rate through the defects from that of air to that of a polymer which improves the overall barrier performance.

8.5 Multilayers

Where there has been a planarising layer and an inorganic coating deposited and the barrier performance still does not meet the desired performance it is possible to repeat the process of depositing alternating organic and inorganic layers.

Keep in mind that it has been shown that it is possible to achieve an ultra barrier performance using just a single layer each of an organic and inorganic coating. Therefore if this ultra barrier performance is not achieved it implies that the cleanliness of the substrate, handling quality or quality of the deposition process is not as good as it could be and the result is that there are still defects present. The farther away from a clean process you are the more layers will need to be deposited to make up for the deficiencies in the materials and processing.

8.6 Conclusion

As you can see from the above all the way through the process the one thing that stands out as being critical to the barrier performance and that is cleanliness. The cleanliness of the substrate surface is critical as any surface defect or particulate contaminant can result in an imperfect coating and loss of barrier performance. From that point onwards any further particulate contamination will result in damage to the coating and a loss in barrier performance.

For the best barrier coatings there is no such thing as too clean.

Index

Author Biography

Dr. Charles A. Bishop
C.Eng., MIMMM

Whilst serving an apprenticeship and qualifying as a toolmaker, Charles gained an ONC & HNC in Mechanical Engineering at the local Technical College. He followed this by taking a degree in Materials Engineering and a Diploma in Industrial Studies. It was during this time he was first exposed to the mysteries of vacuum processing as he helped develop a new implant for prosthetic surgery. Transferring from the Materials to Physics department he completed a Masters and finally a Doctoral degree by research into vacuum processing. Following a period of contract research and consulting he was recruited by ICI New Science Group and later into the Polyester Business where he spent time looking to develop future technologies that might be suitable platforms for new products. Included in this job were secondments into other business areas as diverse as security pigments and vacuum deposited security devices, optical data storage, pyrotechnics for airbag enhancers and antistatic coatings for fibres to name but a few. In 1998 Charles opted to return to being a consultant and started C.A.Bishop Consulting Ltd., offering access to expertise gained from over 25 years experience of vacuum engineering, system and process design, troubleshooting and training.

He is the author of the book *Vacuum deposition onto webs, films and foils* and author or co-author of over 60 papers and

5 patents. Charles has regularly contributed to conferences such as those run by the Society of Vacuum Coaters (SVC) and Association of Industrial Metallizers, Coaters & Laminators (AIMCAL). He was presented with a SVC Mentor Award 2008. Currently he is the Editor of www.vacuumcoatingblog.com on behalf of AIMCAL as well as a regular Lecturer at the AIMCAL Converting Schools and presenter of a number of Webinars also offered by AIMCAL.

Also of Interest

Forthcoming related titles from Scrivener Publishing

Introduction to Surface Engineering and Functionally Engineered Materials by Peter Martin, *President, Columbia Basin Thin Film Solutions LLC, Emeritus Fellow, Pacific Northwest National Laboratory.* Forthcoming summer 2011.

The *Introduction* will provide a clear and understandable resource for users and developers of advanced engineered materials, particularly in the area of thin films.

Surfaced engineered materials are used in a variety of industries including automotive, aerospace, missile, power, electronic, biomedical, textile, petroleum, petrochemical, chemical, steel, power, cement, machine tools, and construction. Surface engineering techniques are used to develop a wide range of functional properties, including physical, optical, chemical, electrical, electronic, magnetic, mechanical, wear-resistant and corrosion-resistant properties at the required substrate surfaces. Engineered materials are now being developed for and used in advanced photovoltaic devices, dye sensitized solar cells, quantum cascade lasers, advanced electronics, drug delivery, medical devices, metamaterials, optical photonic bandgap devices, negative refractive index devices, superlenses, artificial magnetism, cloaking devices, thermoelectric power generation and much more.

This book will introduce the aforementioned techniques and materials so that industry engineers and product developers, as well as students and researchers, will come to grips with the rapidly developing field of surface engineering and engineered materials.

Introduction to Plasmas for Materials Processing by Scott Walton, *Plasma Physics Division, Naval Research Laboratory, Washington, D.C.*
Forthcoming fall 2011.

The aim of this work is to instruct engineers and product developers that use plasma-based processing equipment and provide them with a skill set that empowers them to take a more proactive approach by fully understanding the plasma-surface interface.

The three main areas of interest to any plasma processing device: plasma generation, gas-phase physics and chemistry, and the plasma-surface interface, are not independent of each other, so to fully understand the plasma-surface interface, one must understand all aspects of the device. This book will discuss all three areas although the emphasis will be on the plasma-surface interface.

Cleaning for Semiconductor Manufacturing: Fundamentals and Applications edited by Karen Reinhardt, *Cameo Consulting* and Richard F. Reidy, *Dept. of Materials Science and Engineering, University of North Texas.*
Forthcoming fall 2010. ISBN 978-0-470-62595-8.

The volume provides an in-depth discussion of the fundamentals of cleaning and surface conditioning of semiconductor applications such as high-k/metal gate cleaning, copper/low-k cleaning, high dose implant stripping, and silicon and SiGe passivation. The theory and fundamental physics associated with wet etching and wet cleaning is reviewed, plus the surface and colloidal aspects of wet processing. Formulation development practices and methodology are presented along with the applications for preventing copper corrosion, cleaning aluminum lines, and other sensitive layers. Other chapters are devoted to outlining the analytical methods for best evaluating the cleanliness of the wafer surface. Supporting chapters on filtering and recirculation of the chemicals are included as well.

Polymer Nanotube Nanocomposites: Synthesis, Properties, and Applications Edited by Vikas Mittal, BASF Polymer Research, Germany June 2010. ISBN 978-0-470-62592-7.

The purpose of this edited volume is to bring together contributions from a variety of senior scientists in the field of polymer nanotube composites technology to shed light on the recent advances in these commercially important areas of polymer technology. The book provides the following features:

- Sums up recent advances in nanotube composite synthesis technology

- Provides a basic introduction to polymer nanotubes nanocomposite technology for the readers new to the field
- Provides valuable insights for the use of technologies for polymer nanocomposites for commercial application.